Stem Cells in Clinical Applications

Series Editor
Phuc Van Pham
Laboratory of Stem Cell Research and Application
University of Science, Vietnam National University
Ho Chi Minh City, Vietnam

Stem Cells in Clinical Applications brings some of the field's most renowned scientists and clinicians together with emerging talents and disseminates their cutting-edge clinical research to help shape future therapies. While each book tends to focus on regenerative medicine for a certain organ or system (e.g. Liver, Lung and Heart; Brain and Spinal Cord, etc.) each volume also deals with topics like the safety of stem cell transplantation, evidence for clinical applications, including effects and side effects, guidelines for clinical stem cell manipulation and much more. Volumes will also discuss mesenchymal stem cell transplantation in autoimmune disease treatment, stem cell gene therapy in pre-clinical and clinical contexts, clinical use of stem cells in neurological degenerative disease, and best practices for manufacturers in stem cell production. Later volumes will be devoted to safety, ethics and regulations, stem cell banking and treatment of cancer and genetic disease. This series provides insight not only into novel research in stem cells but also their clinical and real-world contexts. Each book in *Stem Cells in Clinical Applications* is an invaluable resource for advanced undergraduate students, graduate students, researchers and clinicians in Stem Cells, Tissue Engineering, Biomedical Engineering or Regenerative Medicine.

More information about this series at http://www.springer.com/series/14002

Phuc Van Pham • Ahmed El-Hashash
Editors

Stem Cells for Cancer and Genetic Disease Treatment

Springer

Editors
Phuc Van Pham
Laboratory of Stem Cell Research and Application & Stem Cell Institute & Cancer Research Laboratory
VNUHCM University of Science
Ho Chi Minh City, Vietnam

Ahmed El-Hashash
University of Edinburgh - Zhejiang International Campus (UoE-ZJU)
Haining, Zhejiang, China

ISSN 2365-4198 ISSN 2365-4201 (electronic)
Stem Cells in Clinical Applications
ISBN 978-3-319-98064-5 ISBN 978-3-319-98065-2 (eBook)
https://doi.org/10.1007/978-3-319-98065-2

Library of Congress Control Number: 2018959439

This Springer imprint is published by the registered company Springer Nature Switzerland AG
The registered company address is: Gewerbestrasse 11, 6330 Cham, Switzerland

Preface

Stem cell research can be traced back to more than 20 years ago when Evans MJ, Kaufman and Martin GR first isolated embryonic stem cells from mouse blastocysts in 1981and when a research article announcing the discovery of human embryonic stem cells emerged in 1998 by Thomson JA. The stem cell research field has rapidly expanded as new research and experience broadens our knowledge of different aspects of stem cell biology and applications. In the past decade, the stem cell field has grown very rapidly and continues to be one of the most exciting aspects of biomedical research. This book brings together a number of topics that are related to stem cell biology, formation and differentiation, as well as stem cell applications in the treatment of cancer and genetic disorders.

This volume of the *Stem Cells in Clinical Applications* book series, with the title ***Stem cells for Cancer and Genetic Disease Treatment***, aims to provide an invaluable updated resource for advanced undergraduate students, graduate students, researchers and clinicians in some stem cell applications for cancer and genetic disease treatment.

The book has 10 chapters, covering almost all of the present applications of stem cells in cancer and genetic diseases. Chapters **1–6** introduce the role of stem cells for cancer treatment, where applications of cancer stem cells (Chaps. **1** and **4**), mesenchymal stem cells (Chap. **2**) and other progenitor cells in various cancers (Chaps. **3**, **5**, and **6**) are discussed. Chapters **7–10** introduce some applications of stem cells for genetic diseases, including systemic lupus erythematosus (Chap. **7**), cardiomyopathy (Chap. **8**), genetic neurodegenerative disease (Chap. **9**), and nerve and muscle repair (Chap. **10**).

Although we could not hope to be comprehensive in the coverage of stem cells of different tissues, our main aim in compiling this book was to bring together a selection of the current progress in understanding stem cell applications in the treatments of cancer and genetic disorders.

In preparing this book, we aimed at making it accessible not only to those working in the stem cell biology field, but also to non-experts with a broad interest in stem cells and in human health. Our hope is that this book will be of value to all concerned with stem cell application in medicine.

We are indebted to our authors, who graciously accepted their assignments and who have infused the text with their energetic contributions. We are incredibly thankful to the staff at Springer Nature, who published this book.

Ho Chi Minh, Vietnam Phuc Van Pham
Haining, Zhejiang, China Ahmed El-Hashash

Contents

Contributors

Muhammad Sajid Hamid Akash Department of Pharmaceutical Chemistry, Government College University Faisalabad, Faisalabad, Punjab, Pakistan

Babak Arjmand Cell Therapy and Regenerative Medicine Research Center, Endocrinology and Metabolism Molecular-Cellular Sciences Institute, Tehran University of Medical Sciences, Tehran, Iran

Metabolomics and Genomics Research Center, Endocrinology and Metabolism Molecular-Cellular Sciences Institute, Tehran University of Medical Sciences, Tehran, Iran

Ashima Bhan Dr. D. Y. Patil Biotechnology and Bioinformatics Institute, Dr. D. Y. Patil Vidyapeeth, Pune, Maharashtra, India

Ahmed El-Hashash University of Edinburgh-Zhejiang University Institute, Molecular Medicine Research Centre and Centre of Stem Cell and Regenerative Medicine, Schools of Medicine and Basic Medicine, Zhejiang University, Hangzhou, and the University of Edinburgh, Edinburgh, UK

The University of Edinburgh-Zhejiang International Campus, (UoE-ZJU), Arts and Science Building, Haining, Zhejiang, China

Gavino Faa Division of Pathology, Department of Surgical Sciences, University of Cagliari, Cagliari, Italy

Daniela Fanni Division of Pathology, Department of Surgical Sciences, University of Cagliari, Cagliari, Italy

Clara Gerosa Division of Pathology, Department of Surgical Sciences, University of Cagliari, Cagliari, Italy

Parisa Goodarzi Brain and Spinal Cord Injury Research Center, Tehran University of Medical Sciences, Tehran, Iran

Komal Jabeen Institute of Pharmacy, Physiology and Pharmacology, University of Agriculture, Faisalabad, Punjab, Pakistan

Xuexiao Jin Institute of Immunology, Zhejiang University School of Medicine, Hangzhou, Zhejiang, China

Zhejiang University–University of Edinburgh Institute, Zhejiang University School of Medicine, Hangzhou, Zhejiang, China

John Ku University of Edinburgh-Zhejiang University (UoE-ZJU) Biomedicine Institute, Molecular Medicine Research Centre and Centre of Stem Cell and Regenerative Medicine, Schools of Medicine and Basic Medicine, Zhejiang University, Hangzhou, Zhejiang, China

Edinburgh Medical School, University of Edinburgh, Edinburgh, UK

California State Polytechnic University, Pomona, California, USA

Federica Lai Division of Pathology, Department of Surgical Sciences, University of Cagliari, Cagliari, Italy

Bagher Larijani Endocrinology and Metabolism Research Center, Endocrinology and Metabolism Clinical Sciences Institute, Tehran University of Medical Sciences, Tehran, Iran

Shen Lu Veritas Genetics Asia Inc., Hangzhou, Zhejiang, China

Eiman Abdel Meguid Centre for Biomedical Sciences Education, School of Medicine, Dentistry and Biomedical Sciences, Queen's University Belfast, Belfast, Northern Ireland, UK

Fakher Rahim Health Research Institute, Thalassemia and Hemoglobinopathies Research Center, Ahvaz Jundishapur University of Medical Sciences, Ahvaz, Iran

T. K. Rajendra Dr. D. Y. Patil Biotechnology and Bioinformatics Institute, Dr. D. Y. Patil Vidyapeeth, Pune, Maharashtra, India

Kanwal Rehman Institute of Pharmacy, Physiology and Pharmacology, University of Agriculture, Faisalabad, Punjab, Pakistan

Deepak Rohila Institute of Immunology, Zhejiang University School of Medicine, Hangzhou, China

Satish Sasikumar Dr. D. Y. Patil Biotechnology and Bioinformatics Institute, Dr. D. Y. Patil Vidyapeeth, Pune, Maharashtra, India

Phuc Van Pham Laboratory of Stem Cell Research and Application & Stem Cell Institute & Cancer Research Laboratory, VNUHCM University of Science, Ho Chi Minh City, Vietnam

Ngoc Bich Vu Laboratory of Stem Cell Research and Application & Stem Cell Institute, VNUHCM University of Science, Ho Chi Minh City, Vietnam

Yang Yao Institute of Immunology, Zhejiang University School of Medicine, Hangzhou, Zhejiang, China

Part I
Stem Cells for Cancer Treatment

Chapter 1
Cancer Stem Cells and Their Therapeutic Applications

Eiman Abdel Meguid and Ahmed El-Hashash

Abbreviations

AML	Acute myelogenous leukemia
BMPs	Bone morphogenetic proteins
CML	Chronic myelogenous leukemia
CSCs	Cancer stem cells
EMT	Epithelial-mesenchymal transition
HSCs	Hematopoietic stem cells
Sca-1	Stem cell antigen-1
VEGF	Vascular endothelial growth factor

1.1 Introduction

Stem cells are characterized by their proliferation/self-renewal ability to form new stem cells and their ability to differentiate into many specialized cell types. Many research studies have suggested a cellular hierarchy in tumor development, with a small number of cells having the ability of recapitulating the cancer via the processes of tumor initiation and propagation (Quintana et al. 2008). Notably, this small group of highly efficient tumor-initiating cells shows certain characteristics of multipotent stem cells, including the abilities to self-renew and differentiate, and it is called

E. A. Meguid
Centre for Biomedical Sciences Education, School of Medicine, Dentistry and Biomedical Sciences, Queen's University Belfast, Belfast, Northern Ireland, UK

A. El-Hashash (✉)
The University of Edinburgh-Zhejiang International campus (UoE-ZJU Institute), and Centre of Stem Cell and Regenerative Medicine, Schools of Medicine & Basic Medicine, Zhejiang University, Haining, Zhejiang, China

The University of Edinburgh-Zhejiang International Campus, (UoE-ZJU), Arts and Science Building, Haining, Zhejiang, China

P. V. Pham, A. El-Hashash (eds.), *Stem Cells for Cancer and Genetic Disease Treatment*, Stem Cells in Clinical Applications, https://doi.org/10.1007/978-3-319-98065-2_1

cancer stem cells (Tang 2012). For example, stem cells were detected in breast cancer, colorectal and other types of tissue-specific cancers (Al-Hajj et al. 2004; Nassar and Blanpain 2016; Batlle and Clevers 2017; Aponte and Caicedo 2017; Munro et al. 2018). CSCs are tumorigenic, have the metastatic phenotypes, and probably result from adult neoplastic cells with stem cell properties via mutations or produced by specific mutations in normal stem cells (Rich and Bao 2007; Gupta et al. 2011). There is current debate on the functional roles of the niche in the development of CSCs and/or the prevention of tumorigenesis (Aponte and Caicedo 2017).

1.2 The Stem Cell Concept of Cancerogenesis

Early studies on leukemia cells have shown evidences that support the concept of stem cells in cancerogenesis (Lapidot et al. 1994; Bonnet and Dick 1997). These studies detected a small population of cells that are tumorigenic and express certain distinct and characteristic markers (Lapidot et al. 1994; Bonnet and Dick 1997). In addition, this hypothesis indicates that certain cellular components with characteristic properties of stem cells, including self-renewal/proliferation and differentiation, are involved in the formation and derivation of tumors (Kreso et al. 2014; Nassar and Blanpain 2016; Batlle and Clevers 2017; Aponte and Caicedo 2017). These early studies aimed to model leukemic disease in vivo by extracting hematopoietic cells from human patients afflicted with acute myelogenous leukemia (AML) and transplanting them into immunodeficiency mice (Lapidot et al. 1994; Bonnet and Dick 1997; Zhu et al. 2002). Results showed that a core group of cells that express CD34, a cell surface marker primarily associated with hematopoietic stem cells (HSCs), can generate leukemia in xenografts (Lapidot et al. 1994; Bonnet and Dick 1997; Zhu et al. 2002).

The discovery of cancer cells that exhibit certain stem cell properties has also raised the possibility that cancer types other than AML may be linked to and contain CSCs (Bonnet and Dick 1997; Hope et al. 2004). For example, a small population of cells that express the cancer stem cell markers CD24 and CD44 was detected in human breast cancers (Al-Hajj et al. 2003). These CD44+ CD24− cells can grow in culture and form tumor mammospheres (Ponti et al. 2005; Dontu et al. 2003) as well as secrete the vascular endothelial growth factor (VEGF; Ponti et al. 2005).

Similarly, CSCs expressing nestin and CD133 neural stem cell markers present in brain tumors in humans, like their normal counterparts, and can form neurospheres in culture (Singh et al. 2004a, b). In addition, a cell subpopulation that is CD44+/a2h1 hi/CD133+ and has distinct prosperities of stem cells exists in human prostate cancer (Richardson et al. 2004). Furthermore, CSCs were reported in lung cancer (Kim et al. 2005).

There are many evidences, which suggest a role of stem cells in carcinogenesis. For example, there are some common properties that are shared between cancer cells and normal stem cells, including their abilities to express telomerase, migrate, self-renew/proliferate and differentiate, and spread by metastasis, as well as activate the anti-apoptotic regulatory (Gil et al. 2008; Atena et al. 2014; Nassar and Blanpain 2016; Batlle and Clevers 2017; Aponte and Caicedo 2017; Papaccio et al. 2017). In addition, CSCs were hypothesized to exist within solid tumors, particularly in tissues that show a characteristically rapid self-renewal/proliferation capacity (Dalerba et al. 2007). This was evidenced by the isolation of putative CSCs from a breast cancer, leading the detection of a population of cells that are $CD44^{+}CD24^{-}$, which are surface markers of normal mammary stem cells (Al-Hajj et al. 2003). Indeed, re-injection of these CSCs into immuno-deficient mice can lead to breast cancer in eight out of nine cases (Al-Hajj et al. 2003). In contrast, other cell types that are derived from the tumor mass could not result in tumor formation after injection (Al-Hajj et al. 2003; Ponti et al. 2005). In addition, CSCs were implicated in the development of other types of solid tumor cancers such as colon cancer and colorectal cancer (Liu et al. 2009; Ricci-Vitiani et al. 2007). Furthermore, genetic analyses demonstrated a strong correlation between the derived brain tumor stem cells and the original patient tumors (Eyler and Rich 2008; Piccirillo et al. 2006).

1.3 Dysregulation of Stem Cell Self-Renewal and Cancer Transformation

Tumors may arise from self-renewing tissue-specific stem/progenitor cells through a process called cancer transformation. This hypothesis is supported by the finding that the blast crisis, which is the last phase of chronic myelogenous leukemia (CML) development, may originate in hematopoietic progenitors because of dysregulated Wnt signaling pathway that allows their self-renewing, a property that is restricted to HSCs (Jamieson et al. 2004). In addition, transfecting AML-ETO into pure hematopoietic progenitor cell populations can promote both their self-renewing capability and myeloid transformation (Kelly and Gilliland 2002). Similarly, transformations of mammary stem/progenitor cells can lead to the formation of breast cancer in humans (Dontu et al. 2003). Moreover, transformations of HSCs may lead to the formation of leukemic stem cells (Hope et al. 2004).

Both dysregulated process of cell self-renewal and preferential symmetric mode of cell division are major and early events during cancer or tumor transformation (Liu et al. 2005; Atena et al. 2014; Kreso et al. 2014; Batlle and Clevers 2017). Dysregulation of the self-renewal process is well exemplified in models of bone marrow transplantations, in which one HSC is injected into lethally irradiated mice. This can lead to the processes of repopulating stem cell pool and reconstituting the hematopoietic system (Ito and Suda 2014). Dysregulation of cell self-renewal can lead to an inappropriate expansion of stem cell number, leading to a tumor formation (Ito and Suda 2014).

Several signaling pathways regulate the process of self-renewing stem cells, including Notch, Wnt, and Hedgehog pathways, during the normal development of different tissues such as neuronal, hematopoietic, and mammary gland tissues (Atena et al. 2014; Kreso et al. 2014; Nassar and Blanpain 2016; Batlle and Clevers 2017).

Defects in one or more of these signaling pathways can lead to tumorigenesis in animal models of cancer and human cancer (Liu et al. 2005; Dontu et al. 2004; Atena et al. 2014; Kreso et al. 2014; Nassar and Blanpain 2016; Batlle and Clevers 2017). For example, disruption of the Wnt pathway is remarkable during the development of colon cancer (Dontu et al. 2004). Similarly, defects in the Hedgehog pathway activities were detected in the basal carcinoma of human skin (Unden et al. 1996), and in human breast, pancreatic, prostate, and gastric carcinomas (Karhadkar et al. 2004), while defects in Notch signaling activities were shown in both breast and cervical cancers and lymphoblastic leukemia in humans (Dievart et al. 1999; Nam et al. 2002; Nickoloff et al. 2003; Siziopikou et al. 2003; Benson et al. 2004).

1.4 CSCs in Cancer Therapy

CSCs have characteristically more effective resistance to therapeutic treatments than bulk cells that results in the failure of conventional cancer chemotherapeutic approaches. A good example for this effective resistance is in melanoma cancer cells (Hanahan and Weinberg 2011). This probably represents a major challenge to the targeted therapy. In addition, both cancer and non-cancerous stem cells could display protective drug transport mechanisms, which remove cytotoxic chemicals from the cell (Dean et al. 2005; Blagosklonny 2006). Furthermore, CSCs show a characteristically enhanced radiation resistance through either the activation of DNA repair mechanisms or an increased defense against reactive oxygen molecules/species (ROS; Bao et al. 2006; Diehn et al. 2009). Therefore, new and innovative therapeutic approaches are needed for the reversal of chemotherapy resistance and other resistance types of targeted CSCs.

Different therapies of cancer are often limited by their lack of specificity and high incidence of adverse effects. For example, traditional chemotherapeutic methods rely on toxicity to both cancerous and non-cancerous cells since radiation therapy cannot specifically target tumor cells. The stem cell concept of cancerogenesis can help in developing novel cancer therapeutic strategies that can solve the limitation problems of currently available treatments for cancer, by designing new therapies for the tumor-initiating cellular compartment, CSCs.

Eliminating CSCs, while maintaining normal stem cells, is required for effective destructions of cancer in different body organs such as the bone marrow and gut. A major challenge for this strategy is that several signaling pathways controlling stem cell biology and behavior are shared by both CSCs and normal stem cells. However, many research studies have successfully targeted these signaling pathways in animal models of cancer, suggesting the feasibility of this therapeutic approach. For example, a targeted inhibition of Notch signaling activities may suppress the development of breast cancers that overexpress Notch1 (Weijzen et al. 2002; Pece et al. 2004), while an inhibition of the Hedgehog signaling activities can block the formation of medulloblastoma in a transgenic mouse model (Romer et al. 2004).

Cancer therapeutic strategies including both conventional therapies that target bulk tumor cells and the suppression of certain regulatory pathways of CSCs can improve cancer treatments, compared to monotherapies (Linkous and Yazlovitskaya 2012; Baumann et al. 2008; Shigdar et al. 2012). For example, PI3K/AKT signaling is important for the maintenance of prostate tumor-initiating cells, and its targeted suppression can eliminate CD133+/CD44+ tumor stem/progenitor cell populations both in culture and in xenograft tumors (Dubrovska et al. 2009). This targeted suppression can also sensitize prostate tumors in Taxoterein mouse models (Dubrovska et al. 2009). In addition, targeted inhibition of PI3K/AKT in animal models, under hypoxic and normoxic conditions, can suppress the process of epithelial cells transition into mesenchymal cells and radio-sensitize prostate cancer cells (Chang et al. 2013; Potiron et al. 2013).

One potential therapy for cancer is targeting the CSC niche by blocking the factors and signals such as vascular endothelial growth factor (VEGF) that are used by these stem cells to produce and stabilize supporting tissues. The expression level of VEGF increases in tumors, causing angiogenesis that provides the necessary blood supply to the developing tumor (Plate et al. 1992; Kim et al. 1993). One major risk for this kind of therapeutic approaches is the similarities between CSCs and other adult normal stem cells of the treated patient (Reya and Clevers 2005). Since the targeted cell surface markers and factors exist in both types of stem cells (normal and CSCs), these therapeutic methods may be inadequate and non-specific. In addition, they probably have an associated toxicity to other normal tissues.

Another potential therapy based on CSCs is to target the unique markers that are expressed on the surface of these CSCs such as CD44 (Williams and Cancelas 2006). A specific CD44 antibody can be used to target CD44 cell marker that can lead to loss of the tumor cell ability to effectively engraft, and subsequent inhibition of tumor growth and loss of primitive cell types within tumor grafts (Jin et al. 2006). This approach is promising; however, other studies have demonstrated that the treatment with antibodies had little effect on tumor formation in xenograft (Krause et al. 2006). However, a major challenge for this approach is to identify and use the characteristic cell surface makers of CSCs that are not expressed on the surface of normal stem cells.

Nanomedicine is an innovative approach with a great potential to develop novel agents that can provide beneficial pharmacological properties to eliminate CSCs (Shen et al. 2016; Shkembi et al. 2016; Qin et al. 2017). In nanomedicine, the nano-vehicles are used for both transporting various therapeutic drugs and facilitating the process of cellular uptake (Woodward et al. 2007). In addition, the nano-vehicles could help in both prolonging circulation and improving the biodistribution of incorporated drugs as well as encapsulating chemotherapeutics (Woodward et al. 2007; Diehn et al. 2009). Developing nanomedicine for designing different therapeutic strategies for CSCs requires both harnessing the potential of nanomedicine and nanobiology, and better understanding of the properties of CSCs and their roles in the progression of cancer.

Differentiation cancer therapies are also important for CSCs. In different cancer types, CSCs have several characteristic and distinguishable metabolic features compared to their normal counterparts, including high glycolytic activity and low mitochondrial respiration (Pecqueur et al. 2013). Targeting some key metabolic enzymes such as PGK1 can stimulate tumor stem cell differentiation in gastric cancers (Pecqueur et al. 2013). In addition, several studies suggest the use of bone morphogenetic proteins (BMPs) for stem cell differentiation therapies of glioblastoma, while ATRA can be used for leukemia and some solid tumors (Formelli and Cleris 1993; Fenaux et al. 1999; Piccirillo et al. 2006; Zieker et al. 2013). Furthermore, other cancer therapeutic approaches that rely on the eradication of CSCs by modifying the tumor microenvironment through re-oxygenation of CSC niche have a great potential for application (Reisfeld 2013; Yang et al. 2014).

1.5 Conclusions and Future Directions

Cancer is currently a major health problem worldwide. CSC hypothesis is well supported by several studies that detected a small population of cells that are tumorigenic, express certain distinct and characteristic markers, and show characteristic properties of stem cells, including self-renewal/proliferation and differentiation, suggesting a role of stem cells in carcinogenesis. Better understanding and characterization of CSCs can help in devising more effective therapeutic strategies and approaches for cancer. More research is, however, still needed on the biology, behavior, and environment of these CSCs since this will help in improving their applications in cancer therapies.

One major challenge for CSC researches and tumor/cancer therapy is the lack of specific and unique CSC markers. Most studies detect CSCs in vivo or in vitro using common markers, including CD44 and CD133 that are also expressed on the surface of normal stem cells in adult humans (Al-Hajj et al. 2003; Singh et al. 2004a, b; Liu et al. 2005). Targeted therapies directed against these markers will, therefore, likely ablate both CSC and normal stem cell populations. Therefore, identification and validation of specific CSCs markers and biomarkers are currently required for effective tumor diagnostics and making a substantial progress in cancer therapies.

Another major challenge is optimizing protocols for more effective isolation of CSCs. Isolated CSCs could be screened for molecular defects before sensitization to specific drugs or subjecting them to a direct drug screen (Gal et al. 2007; Pollard et al. 2009). Yet, more research is still needed to improve and optimize these CSC isolation methods, as well as to establish more feasible and effective protocols for clinical applications. This will help in achieving more effective targeted cancer therapies. In addition, targeted CSC therapies are particularly important for cases in which chemotherapeutics are practically ineffective, including hepatic carcinoma with a median response below 15% (Cho et al. 2008).

Furthermore, the characterization of resistance/adaptive mechanisms of CSCs to currently available treatment regimens represents a major obstacle for developing more efficient cancer therapies. Molecular-targeted therapeutic approaches have shown early promises in CSCs and cancer treatments (Soeda et al. 2008; Hambardzumyan et al. 2008). For example, the block of specific resistance mechanisms of CSCs to traditional therapies, including IL-4 blockade, was successful in sensitizing colon CSCs to chemotherapy (Francipane et al. 2008). However, CSCs are likely a dynamic cell population through the process of adaptive/resistance responses or clonal evolution. More research is still required in this important area.

Despite the heterogeneity of cancer, more advancements in cancer diagnostic methods can help to establish an effective customized medicine for individual cancer patients. In addition, future improvements of genetic arrays will help in clarifying the molecular variations among various types of human cancer, while more advanced progresses in genetic signatures will help in predicting cancer outcome and potential treatment methods instead of only using traditional histological analyses.

In summary, more therapeutic innovations are still needed for cancer. Better characterization and understanding of CSC biology, behavior, and environment that are still largely uncovered can lead to the development of these therapeutic innovations for cancer. In addition, identification and characterization of major signaling pathways and factors that regulate CSC biology and behavior, including TGFβ, Notch, Wnt, and Hedgehog pathways, may enable the development of more efficient drug targeting therapies. Furthermore, better characterizations of other regulatory mechanisms of CSC pathology or mechanisms controlling CSC interactions with their niche and mechanisms of cancer metastasis will eventually lead to the improvement of cancer treatments.

Conflict of Interest The author confirms that this book contents have no conflict of interest.

References

Al-Hajj M, Wicha MS, Benito-Hernandez A, Morrison SJ, Clarke MF (2003) Prospective identification of tumorigenic breast cancer cells. Proc Natl Acad Sci U S A 100:3983–3988

Al-Hajj M, Becker MW, Wicha M, Weissman IL, Clarke MF (2004) Therapeutic implications of cancer stem cells. Curr Opin Genet Dev 14:43–47

Aponte PM, Caicedo A (2017) Stemness in cancer: stem cells, cancer stem cells, and their microenvironment. Stem Cells Intl 2017:1–17

Atena M, Reza AM, Mehran G (2014) A review on the biology of cancer stem cells. Stem Cell Discov 4:83–89

Batlle E, Clevers H (2017) Cancer stem cells revisited. Nat Med 23:1124–1134

Baumann M, Krause M, Hill R (2008) Exploring the role of cancer stem cells in radioresistance. Nat Rev Cancer 8:545–554

Benson RA, Lowrey JA, Lamb JR, Howie SE (2004) The Notch and Sonic hedgehog signalling pathways in immunity. Mol Immunol 41:715–725

Blagosklonny MV (2006) Target for cancer therapy: proliferating cells or stem cells. Leukemia 20:385–391

Bonnet D, Dick JE (1997) Human acute myeloid leukemia is organized as a hierarchy that originates from a primitive hematopoietic cell. Nat Med 3:730–737

Bao S, Wu Q, Sathornsumetee S et al (2006 August 15) Stem cell-like glioma cells promote tumor angiogenesis through vascular endothelial growth factor. Cancer Res 66(16):7843–7848

Chang L, Graham PH, Hao J, Ni J, Bucci J, Cozzi PJ et al (2013) Acquisition of epithelial–mesenchymal transition and cancer stem cell phenotypes is associated with activation of the PI3K/Akt/mTOR pathway in prostate cancer radioresistance. Cell Death Dis 4:875–884
Cho K, Wang X, Nie S, Chen ZG, Shin DM (2008) Therapeutic nanoparticles for drug delivery in cancer. Clin Cancer Res 14:1310–1316
Dalerba P, Cho RW, Clarke MF (2007) Cancer stem cells: models and concepts. Annu Rev Med 58:267–284
Dean M, Fojo T, Bates S (2005) Tumour stem cells and drug resistance. Nat Rev Cancer 5:275–284
Diehn M, Cho RW, Lobo NA, Kalisky T, Dorie MJ, Kulp AN et al (2009) Association of reactive oxygen species levels and radioresistance in cancer stem cells. Nature 458:780–783
Dievart A, Beaulieu N, Jolicoeur P (1999) Involvement of Notch1 in the development of mouse mammary tumors. Oncogene 18:5973–5981
Dontu G, Al-Hajj M, Abdallah WM, Clarke MF, Wicha MS (2003) Stem cells in normal breast development and breast cancer. Cell Prolif 36:59–72
Dontu G, Jackson KW, McNicholas E, Kawamura MJ, Abdallah WM, Wicha MS (2004) Role of Notch signaling in cell-fate determination of human mammary stem/progenitor cells. Breast Cancer Res 6:605–615
Dubrovska A, Kim S, Salamone RJ, Walker JR, Maira S-M, García-Echeverría C et al (2009) The role of PTEN/Akt/PI3K signaling in the maintenance and viability of prostate cancer stem-like cell populations. Proc Natl Acad Sci U S A 106(1):268–273
Eyler CE, Rich JN (2008) Survival of the fittest: cancer stem cells in therapeutic resistance and angiogenesis. J Clin Oncol 26:2839–2845
Fenaux P, Chastang C, Chevret S, Sanz M, Dombret H, Archimbaud E et al (1999) A randomized comparison of all transretinoic acid (ATRA) followed by chemotherapy and ATRA plus chemotherapy and the role of maintenance therapy in newly diagnosed acute promyelocytic leukemia. The European APL Group. Blood 94:1192–1200
Formelli F, Cleris L (1993) Synthetic retinoid fenretinide is effective against a human ovarian carcinoma xenograft and potentiates cisplatin activity. Cancer Res 53:5374–5376
Francipane MG, Alea MP, Lombardo Y, Todaro M, Medema JP, Stassi G (2008) Crucial role of interleukin-4 in the survival of colon cancer stem cells. Cancer Res 68:4022–4025
Gal H, Makovitzki A, Amariglio N, Rechavi G, Ram Z, Givol D (2007) A rapid assay for drug sensitivity of glioblastoma stem cells. Biochem Biophys Res Commun 358:908–913
Gil J, Stembalska A, Pesz KA, Sasiadek MM (2008) Cancer stem cells: the theory and perspectives in cancer therapy. J App Genet 49:193–199
Gupta PB, Fillmore CM, Jiang G, Shapira SD, Tao K, Kuperwasser C et al (2011) Stochastic state transitions give rise to phenotypic equilibrium in populations of cancer cells. Cell 146:633–644
Hambardzumyan D, Becher OJ, Rosenblum MK, Pandolfi PP, Manova-Todorova K, Holland EC (2008) PI3K pathway regulates survival of cancer stem cells residing in the perivascular niche following radiation in medulloblastoma *in vivo*. Genes Dev 22:436–448
Hanahan D, Weinberg RA (2011) Hallmarks of cancer: the next generation. Cell 144:646–674
Hope KJ, Jin L, Dick JE (2004) Acute myeloid leukemia originates from a hierarchy of leukemic stem cell classes that differ in self-renewal capacity. Nat Immunol 5:738–743
Ito K, Suda T (2014) Metabolic requirements for the maintenance of self-renewing stem cells. Nat Rev Mol Cell Biol 15(4):243–256
Jamieson C, Ailles L, Dylla S et al (2004) Granulocyte macrophage progenitors as candidate leukemic stem cells in blast-crisis CML. N Engl J Med 351:657–667
Jin L, Hope KJ, Zhai Q, Smadja-Joffe F, Dick JE (2006) Targeting of CD44 eradicates human acute myeloid leukemic stem cells. Nat Med 12:1167–1174
Karhadkar SS, Bova GS, Abdallah N et al (2004) Hedgehog signalling in prostate regeneration, neoplasia and metastasis. Nature 431:707–712
Kelly LM, Gilliland DG (2002) Genetics of myeloid leukemias. Annu Rev Genomics Hum Genet 3:179–198
Kim KJ, Li B, Winer J, Armanini M, Gillett N, Phillips HS et al (1993) Inhibition of vascular endothelial growth factor-induced angiogenesis suppresses tumour growth *in vivo*. Nature 362:841–844

Kim CF, Jackson EL, Woolfenden AE et al (2005) Identification of bronchioalveolar stem cells in normal lung and lung cancer. Cell 121:823–835
Krause DS, Lazarides K, von Andrian UH, Van Etten RA (2006) Requirement for CD44 in homing and engraftment of BCR-ABL-expressing leukemic stem cells. Nat Med 12:1175–1180
Kreso A et al (2014) Evolution of the cancer stem cell model. Cell Stem Cell 14(3):275–291
Lapidot T, Sirard C, Vormoor J, Murdoch B, Hoang T, Caceres-Cortes J et al (1994) A cell initiating human acute myeloid leukaemia after transplantation into SCID mice. Nature 367:645–648
Linkous AG, Yazlovitskaya EM (2012) Novel radiosensitizing anticancer therapeutics. Anticancer Res 32:2487–2499
Liu S, Dontu G, Wicha MS (2005) Mammary stem cells, selfrenewal pathways, and carcinogenesis. Breast Cancer Res 7:86–95
Liu C, Zhao G, Liu J et al (2009) Novel biodegradable lipid nano complex for siRNA delivery significantly improving the chemosensitivity of human colon cancer stem cells to paclitaxel. J Control Release 140:277–283
Munro MJ, Wickremesekera SK, Peng L et al (2018) Cancer stem cells in colorectal cancer: a review. J Clin Pathol 71:110–116
Nam Y, Aster JC, Blacklow SC (2002) Notch signaling as a therapeutic target. Curr Opin Chem Biol 6:501–509
Nassar D, Blanpain C (2016) Cancer stem cells: basic concepts and therapeutic implications. Annu Rev Pathol 11:47–76
Nickoloff BJ, Osborne BA, Miele L (2003) Notch signaling as a therapeutic target in cancer: a new approach to the development of cell fate modifying agents. Oncogene 22:6598–6608
Papaccio F, Paino F, Regad T, Papaccio G, Desiderio V, Tirino V (2017) Concise review: cancer cells, cancer stem cells, and mesenchymal stem cells: influence in cancer development. Stem Cells Transl Med 6(12):2115–2125
Pece S, Serresi M, Santolini E et al (2004) Loss of negative regulation by Numb over Notch is relevant to human breast carcinogenesis. J Cell Biol 167:215–221
Pecqueur C, Oliver L, Oizel K, Lalier L, Vallette FM (2013) Targeting metabolism to induce cell death in cancer cells and cancer stem cells. Int J Cell Biol 2013:805975
Piccirillo SG, Reynolds BA, Zanetti N, Lamorte G, Binda E, Broggi G et al (2006) Bone morphogenetic proteins inhibit the tumorigenic potential of human brain tumour-initiating cells. Nature 444:761–765
Plate KH, Breier G, Weich HA, Risau W (1992) Vascular endothelial growth factor is a potential tumour angiogenesis factor in human gliomas in vivo. Nature 359:845–848
Pollard SM, Yoshikawa K, Clarke ID, Danovi D, Stricker S, Russell R et al (2009) Glioma stem cell lines expanded in adherent culture have tumor-specific phenotypes and are suitable for chemical and genetic screens. Cell Stem Cell 4:568–580
Ponti D, Costa A, Zaffaroni N, Pratesi G, Petrangolini G, Coradini D, Pilotti S, Pierotti MA, Daidone MG (2005) Isolation and in vitro propagation of tumorigenic breast cancer cells with stem/progenitor cell properties. Cancer Res 65:5506–5511
Potiron VA et al (2013) Radiosensitization of prostate cancer cells by the dual PI3K/mTOR inhibitor BEZ235 under normoxic and hypoxic conditions. Radiother Oncol 106:138–146
Qin W, Huang G, Chen Z, Zhang Y (2017) Nanomaterials in targeting cancer stem cells for cancer therapy. Front Pharmacol 8:1–15
Quintana E, Shackleton M, Sabel MS, Fullen DR, Johnson TM, Morrison SJ (2008) Efficient tumour formation by single human melanoma cells. Nature 456:593–598
Reisfeld RA (2013) The tumor microenvironment: a target for combination therapy of breast cancer. Crit Rev Oncog 18:115–133
Ricci-Vitiani L, Lombardi DG, Pilozzi E et al (2007) Identification and expansion of human colon-cancer-initiating cells. Nature 445:111–115
Rich JN, Bao S (2007) Chemotherapy and cancer stem cells. Cell Stem Cell 1:353–355
Richardson GD, Robson CN, Lang SH, Neal DE, Maitland NJ, Collins AT (2004) CD133, a novel marker for human prostatic epithelial stem cells. J Cell Sci 117:1539–1545
Romer JT, Kimura H, Magdaleno S et al (2004) Suppression of the Shh pathway using a small molecule inhibitor eliminates medulloblastoma in Ptc1(+/_)p53(_/_) mice. Cancer Cell 6:229–240

Reya T, Clevers H (2005) Wnt signalling in stem cells and cancer. Nature 434:843–850. https://doi.org/10.1038/nature03319
Shen S, Xia J-X, Wang J (2016) Nanomedicine-mediated cancer stem cell therapy. Biomaterials 74:1e18
Shigdar S, Lin J, Li Y, Yang CJ, Wei M, Zhus Y et al (2012) Cancer stem cell targeting: the next generation of cancer therapy and molecular imaging. Ther Deliv 3:227–244
Shkembi E, Daniele N, Zinno F, Omar GE (2016) Cancer stem cells and nanomedicine. Peertechz J Cytol Pathol 1(1):048–053
Singh SK, Clarke ID, Hide T, Dirks PB (2004a) Cancer stem cells in nervous system tumors. Oncogene 23:7267–7273
Singh S, Hawkins C, Clarke I et al (2004b) Identification of human brain tumour initiating cells. Nature 432:396–401
Siziopikou K, Miao H, Rizzo P et al (2003) Notch signaling is a therapeutic target in breast cancer. In: Proceedings of the 94th annual meeting of the AACR. pp 1277–1278
Soeda A, Inagaki A, Oka N, Ikegame Y, Aoki H, Yoshimura S et al (2008) Epidermal growth factor plays a crucial role in mitogenic regulation of human brain tumor stem cells. J Biol Chem 283:10958–10966
Tang DG (2012) Understanding cancer stem cell heterogeneity and plasticity. Cell Res 22:457–472
Unden AB, Holmberg E, Lundh-Rozell B et al (1996) Mutations in the human homologue of Drosophila patched (PTCH) in basal cell carcinomas and the Gorlin syndrome: different in vivo mechanisms of PTCH inactivation. Cancer Res 56:4562–4565
Weijzen S, Rizzo P, Braid M et al (2002) Activation of Notch-1 signaling maintains the neoplastic phenotype in human Ras-transformed cells. Nat Med 8:979–986
Williams DA, Cancelas JA (2006) Leukaemia: niche retreats for stem cells. Nature 444:827–828
Woodward WA, Chen MS, Behbod F, Alfaro MP, Buchholz TA, Rosen JM (2007) WNT/beta-catenin mediates radiation resistance of mouse mammary progenitor cells. Proc Natl Acad Sci U S A 104:618–623
Yang X, Zhang X, Fu ML, Weichselbaum RR, Gajewski TF, Guo Y et al (2014) Targeting the tumor microenvironment with interferon-bridges innate and adaptive immune responses. Cancer Cell 25:37–48
Zhu Y, Ghosh P, Charnay P, Burns DK, Prada LF (2002) Nuerofibromas in NF1. Schwann cell origin and role of tumor environment. Science 296:920–922
Zieker D, Bühler S, Ustündag Z, Königsrainer I, Manncke S, Bajaeifer K et al (2013) Induction of tumor stem cell differentiation—novel strategy to overcome therapy resistance in gastric cancer. Langenbecks Arch Surg Dtsch Ges Für Chir 398:603–608

Chapter 2
Mesenchymal Stem Cells as Vectors for Cancer Therapy

Phuc Van Pham and Ngoc Bich Vu

Abbreviations

5-FU	5-fluorouracil
aaTSP-1	Protein thrombospondin
ALL	Acute lymphoblastic leukemia
bFGF	Basic fibroblast growth factor
EGF	Epidermal growth factor
EVs	Extracellular vesicles
EWS	Ewing sarcoma
GFP	Green fluorescent protein
HGF	Hepatic growth factor
hNSCs	Human neural stem cells
HSV-TK	Herpes simplex virus-1 thymidine kinase
HUVEC	Human umbilical cord vein endothelial cells
IFN-β	Interferon beta
IL	Interleukin
MIF	Migration inhibitory factor
MSCs	Mesenchymal stem cells

P. Van Pham (✉)
Laboratory of Stem Cell Research and Application & Stem Cell Institute & Cancer Research Laboratory, VNUHCM University of Science, Ho Chi Minh City, Vietnam
e-mail: pvphuc@hcmuns.edu.vn; phucpham@sci.edu.vn

N. B. Vu
Laboratory of Stem Cell Research and Application & Stem Cell Institute, VNUHCM University of Science, Ho Chi Minh City, Vietnam

P. V. Pham, A. El-Hashash (eds.), *Stem Cells for Cancer and Genetic Disease Treatment*, Stem Cells in Clinical Applications, https://doi.org/10.1007/978-3-319-98065-2_2

NDV	Newcastle disease virus
SDF	Stromal derived factor
UC	Umbilical cord
VEGF	Vascular endothelial growth factor

2.1 Introduction

Cancer is one of the most common causes of death worldwide. In the USA, in 2016, an estimated 1,685,210 new cases of cancer were diagnosed and estimated 595,690 mortalities from the disease (Siegel et al. 2017). Although many methods and drugs—from chemical therapy to radiation therapy and surgery—have been developed which showed some efficacy against cancer, they are still not completely effective against cancer due to tumor evolution and escape mechanisms.

Over the past decade, stem cell therapy has served as a novel treatment strategy for cancer. In general, there are three strategies to use stem cells for cancer treatment. Firstly, the use of hematopoietic stem cells for transplantation in malignant hematopoietic diseases or solid tumors is a method to replace blood cells or to regenerate the hematopoietic system that was destroyed or affected by chemicals or radiation. Secondly, mesenchymal stem cells can be used to carry and deliver anti-tumor agents to tumors. Finally, stem cells can be differentiated into immune cells and then transfused into patients.

In these strategies, the use of mesenchymal stem cells to deliver anti-tumor agents has indicated promising results. This chapter aims to introduce and update the latest results of using engineered MSCs with anti-tumor agents in cancer treatment.

2.2 Mesenchymal Stem Cells and Resources

MSCs are the most common stem cells in the human body. The first description of MSCs was reported in the 1970s by Friedenstein et al. (1970). To define this kind of stem cell, the International Society for Cellular Therapy (ISCT) has suggested some minimal criteria, including: ability to adhere to plastic surface, expression of some MSC markers (e.g., CD44, CD73, CD90, CD105), lack of certain specific markers (including CD14, CD11a, CD34, CD45, CD19, CD79 alpha, and/or HLA-DR). MSCs can be differentiated in vitro into adipocytes, osteoblasts, and chondrocytes (Dominici et al. 2006). During initial discovery, MSCs were determined to be a small population in the bone marrow. To date, they have been identified in various tissues—from fetal tissues to adult tissues. Although these kinds of MSCs express some common markers and characteristics of MSCs, their main characteristic is their secretory profile.

MSCs can easily be isolated and expanded in vitro in common media (e.g., DMEM/F12, DMEM, IMDM) containing 10% fetal bovine serum with or without epidermal growth factor (EGF) and basic fibroblast growth factor (bFGF). Recently some commercial products for MSC culture have also been used to culture and expand MSCs with high efficacy, especially some defined media as well as serum-free media. The achievements of MSC isolation and culture have facilitated the applications of MSCs in the clinic.

The most important characteristic of MSCs for clinical application is low immunogenic expression on their surface. Indeed, allogenic MSCs cause weak immune responses in patients who are recipients of transplantation without HLA matching. Many studies have confirmed that MSCs express low levels of MHC class I, CD80, and CD86 protein, and lack MHC class II.

Given the characteristics of MSCs, commercial companies have developed "off-the-shelf" MSC products or stem cell drugs for treatment of certain diseases. In this strategy, MSCs are collected and expanded from healthy donors, and stored until use in patients. By this strategy, some products were then developed and approved in certain countries; these products include Prochymal (in Canada), HS TemCell (in Japan), and Cartistem (in Korea).

2.3 The Migration of MSCs to Tumors

The migration of MSCs towards tumors is one of the unique properties of MSCs (D'souza et al. 2012; Shah 2013). This phenomenon was observed in several different cancers, including sarcoma (Khakoo et al. 2006), melanoma (Studeny et al. 2002), ovarian cancer (Serakinci et al. 2011), glioma (Sasportas et al. 2009), breast carcinoma (Dwyer et al. 2011), lung cancer (Kanehira et al. 2007), malignant mesothelioma (Sage et al. 2014), pancreatic cancer (Zischek et al. 2009), colon cancer (Zielske et al. 2009), and hepatocellular carcinoma (Li et al. 2010). The mechanism of this process is still unclear. Some recent studies suggest that this migration relates to the paracrine signals from the tumor microenvironment which affects MSCs. Indeed, tumor cells can produce TNF-alpha, interleukins, macrophage migration inhibitory factor (MIF), and stromal derived factor (SDF) at high concentrations. These factors will interact with some receptors on the surface of MSCs (e.g., CXCR4 for SDF-1, c-Met for SCF-c-kit, VEGFR for VEGF, PDGFR for PDGF, CCR2 for MCP-1, and RAGE for HMGB1). These interactions pull MSCs to the tumor environment. Moreover, conditioned medium from tumor cells can upregulate the expression of some receptors on MSCs, especially CXCR4 (Son et al. 2006). MSCs also can migrate to the sites of irradiation or tissue injury (Francois et al. 2006) (Fig. 2.1).

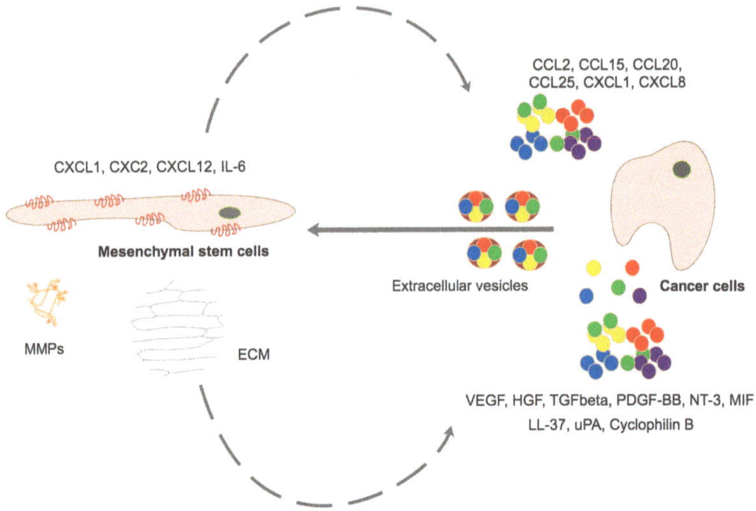

Fig. 2.1 Some signals of tumor cells which attract MSCs. Tumor cells can produce some chemokines such as CCL2, CCL15, CCL20, CCL25, CXCL1, and CXCL8 that can attract MSCs to migrate to tumors via the chemokine receptors that express in the MSCs. Tumor cells also can produce other chemokines as single molecules or in the extracellular vesicles (EVs) such as VEGF, HGF, TGFbeta, PDGF-BB, NT-3, MIF or factors LL-37, uPA and cyclophilin B. Under the tumor environment, MSCs also produce some cytokines (CXCL1, CXCL2, CXCL12, IL-6, and MMPs) to degrade the ECMs and start the migration

2.4 Factors and Agents in MSC Delivery

The natural migration of MSCs to tumors makes them highly suitable vectors for delivery of certain factors or agents to kill tumor cells. MSCs were modified to deliver various factors, such as interleukins, interferons, prodrugs, oncolytic viruses, anti-angiogenic agents, and pro-apoptotic proteins (Table 2.1).

2.4.1 Interleukins

Interleukins are cytokines that can have an important role as anti-tumor factors. Indeed, in human body, helper CD4 T lymphocytes, as well as monocytes, macrophages, and endothelial cells, are all capable of producing interleukins. The interleukins help promote the development and differentiation of T and B lymphocytes, as well as hematopoietic cells.

In this context, to date, MSCs have been modified to express different interleukins (e.g., IL-12, IL-18, and IL-7), and CX3CL1 (Chen et al. 2008; Duan et al. 2009; Gao et al. 2010; Han et al. 2014; Li et al. 2015; Ryu et al. 2011). Modified MSCs produce certain interleukins after transplantation into the recipients. These ILs can improve the anti-cancer surveillance by activating cytotoxic lymphocytes as

Table 2.1 Some factors and agents evaluated for MSC-mediated delivery

Kinds	Factors/agents	References
Interleukins	IL-12	Chen et al. (2008), Duan et al. (2009), Gao et al. (2010), Han et al. (2014), Li et al. (2015), Ryu et al. (2011)
	IL-18	Liu et al. (2015), Sun et al. (2014a), Xu et al. (2015, 2009)
	IL-7	
	CXCL1	
Interferon	IFN-alpha	
	IFN-beta	Chawla-Sarkar et al. (2001), Johns et al. (1992), Ling et al. (2010), Nakamizo et al. (2005), Ren et al. (2008)
	IFN-gamma	Du et al. (2012), Yang et al. (2014)
Enzymes converting prodrugs	Cytosine deaminase (CD)	Cavarretta et al. (2010), Krasikova et al. (2015), Kucerova et al. (2008, 2014)
	Herpes simplex virus (HSV)-1 thymidine kinase (TK)	Kim et al. (2013), Tang et al. (2007)
	Carboxylesterase	
Oncolytic viruses	Newcastle disease virus	Kazimirsky et al. (2016)
	Newcastle disease virus	Castleton et al. (2014)
Anti-angiogenic agents	Protein thrombospondin (aaTSP-1)	van Eekelen et al. (2010)
	Alpha-1-antitrypsin	Ghaedi et al. (2011)
	Pigment epithelium-derived factor	Wang et al. (2013)
	Soluble fms-like tyrosine kinase-1 (sFlt-1)	Li et al. (2017)
Pro-apoptotic proteins	TRAIL	Guiho et al. (2016)
Growth factor antagonists	NK4, an HGF antagonist	Cai et al. (2017), Kanehira et al. (2007), Sun et al. (2014b), Zhu et al. (2014)

well as natural killer cells. In the Chen et al. study (Chen et al. 2008), the authors developed MSCs expressing IL-12 to attack B16 melanoma, 4T1 breast tumor, and HCA hepatoma in mouse models. The results showed that transplantation of IL-12 engineered MSCs in mice showed no systemic toxicity and increased tumor apoptosis ($p < 0.05$) (Chen et al. 2008). In a recent study by Li et al. (2015), adipose-derived stem cells were modified to express IL-12 and then used to treat lung cancer in vitro; the results showed that IL-12 expressing adipose-derived stem cells (ADSCs) successfully inhibited lung adenocarcinoma cell proliferation, migration, and invasion (Li et al. 2015). Moreover, MSCs expressing IL-12 also inhibited malignant ascites in mice (Han et al. 2014), inhibited tumor growth and prolonged survival of glioma-bearing mice (Ryu et al. 2011), reduced the growth of 786-0 renal cell carcinoma (RCC), significantly prolonged RCC mouse survival (Gao et al. 2010), and inhibited Ewing sarcoma tumors (Duan et al. 2009).

IL-18 secreting MSCs also enhanced immune responses and suppressed tumor development. According to Xu et al. (2015) who co-expressed IL-18 and IFN-beta in bone marrow derived mesenchymal stem cells (BM-MSCs) to treat glioma in rats (Xu et al. 2015), the results showed that IL-18 and IFN-beta expressing MSCs inhibited glioma cell growth in vitro, inhibited tumor growth in vivo, and prolonged rat survival (Xu et al. 2009, 2015). Furthermore, umbilical cord derived MSCs (UC-MSCs) expressing IL-18 could suppress the growth, migration, and invasion of breast cancer cells in vitro (Liu et al. 2015; Sun et al. 2014a).

2.4.2 Interferons

IFN-alpha, -beta, and -gamma were transfected into MSCs to elicit immune responses necessary to combat cancers. IFN-beta is the most popular factor that has been used to modify MSCs. IFN-beta expressing MSCs inhibited primary breast tumor growth through inhibition of Stat3 signaling, thereby dramatically reducing pulmonary and hepatic metastases (Ling et al. 2010), and growth of melanoma cell lines (Chawla-Sarkar et al. 2001; Johns et al. 1992). In a study by Nakamizo et al. (2005), IFN-beta modified MSCs were shown to inhibit glioma growth (Nakamizo et al. 2005). The study showed that IFN-beta modified MSCs injected into the cerebral hemisphere significantly increased animal survival. In another study, Ren et al. (2008) expressed IFN-beta in MSCs using recombinant adeno-associated virus (rAAV) and the cells were injected into mice with lung metastasis. The results confirmed that IFN-beta expressing MSCs could reduce tumor volume in lungs and increase natural killer cell activity (Ren et al. 2008).

IFN-gamma modified MSCs also inhibited growth and progression of lung carcinoma via TRAIL (Du et al. 2012; Yang et al. 2014). IFN-alpha expressing MSCs were also evaluated for their anti-tumor activity. Recent studies have shown that transplantation of IFN-alpha-expressing MSCs can reduce the growth of melanoma cells and significantly prolong survival (Ren et al. 2008).

2.4.3 Prodrugs

Prodrugs are non-toxic drugs that can be converted into toxic metabolites, such as cytosine deaminase (CD), herpes simplex virus (HSV)-1 thymidine kinase (TK), and carboxylesterase genes, which confer sensitivity to 5-fluorocytosine (5-FC), ganciclovir (GCV), and camptothecin-11 (CPT-11), respectively. MSCs can be made to express these specific enzymes which convert non-toxic drugs into toxic metabolites. For example, in a study by Kucerova et al. (2008), MSCs were modified to express CD which converts prodrug 5-FC to the drug 5-fluorouracil (5-FU); ADSCs were transfected with CD (CD-ADSCs). The authors injected both

CD-ADSCs and 5-FC into mice bearing xenograft A375 melanoma tumors. Complete regression in 89% of tumor was achieved when 20% CD y-AT-MSC/5-FC were injected (Kucerova et al. 2008, 2014). Inhibition of tumor prostate cancer growth was achieved after intravenous administration of CD y-AT-MSC/5-FC (Cavarretta et al. 2010). Moreover, tumor volume was reduced by 53–58% in mice treated with CD-MSCs, as compared with control mice implanted with B16F10 melanoma (Krasikova et al. 2015). MSCs secreting CD exerted anti-tumor effects on brain tumors when co-injected with 5-FC in animal models. Most importantly, the results confirmed that these therapies were not associated with any detectable adverse effects (Altanerova et al. 2012).

Besides CD, herpes simplex virus thymidine kinase (HSV-TK) is another popular enzyme used to engineer MSCs. HSV-TK converts non-toxic prodrug GCV to a cytotoxic metabolite. Tang et al. (2007) modified ADSCs to express the HSV-TK gene to treat glioblastoma (GBM) in mice. In vitro, cell death of U-87 glioblastoma cells was induced after cells were co-cultured with AT-MSC-HSV-TK in the presence of 25 or 50 μM GCV. In vivo, co-injection of AT-MSC-HSV-TK 1 week after U-87 injection and GCV dramatically reduced tumor size. Particularly, immunohistochemical analyses indicated the presence of AT-MSC-HSV-TK cells in the tumors (Tang et al. 2007). In a recent study, Kim et al. (2013) co-expressed both TRAIL and HSV-TK, or TRAIL or HSV-SV alone, in MSCs. The authors showed that MSCs transfected with both TRAIL and HSV-TK were stronger than those transfected with TRAIL or HSV-TK alone, in response to renal cancer cells. After injection with GCV, administration of MSCs expressing both TRAIL and HSV-TK decreased the number of tumor nodules in the lung (Kim et al. 2013).

2.4.4 *Oncolytic Viruses*

Oncolytic viruses are viruses that can kill neoplastic cells after injection. Therefore, oncolytic viruses (OV) are considered to be promising agents for cancer treatment as they can lyse cancer cells. However, some results using OVs have shown that treatment with OV can be limited by host defense mechanisms and can have harmful effects on some non-cancerous tissues (via the bloodstream). MSCs can host OV replication and transport OVs to tumors; thus, with OVs the treatment efficacy can increase while reducing side effects.

In 2016, Kazimirsky et al. used Newcastle disease virus (NDV), a kind of OV, to infect MSCs to produce NDV-infected MSCs. The conditioned medium from NDV-infected MSCs can induce apoptosis of glioma tumor cells; moreover, conditioned media made glioma stem cells more sensitive to radiation than untreated cells (Kazimirsky et al. 2016). In another study, MSCs carrying OVs were used to successfully treat B-lineage-ALL (acute lymphoblastic leukemia) mice (Castleton et al. 2014).

2.4.5 Anti-angiogenic Agents

Angiogenic agents play an essential role in triggering neoangiogenesis which aids in tumor growth and proliferation. Some studies have suggested that inhibition of angiogenesis can suppress tumor growth. Indeed, in clinical studies, some anti-angiogenic drugs could reduce tumor-associated vasogenic brain edema and resulted in clinical benefit for most patients (Batchelor et al. 2007).

The first study wherein human neural stem cells (hNSCs) were transfected with anti-angiogenic protein thrombospondin (aaTSP-1) was performed in 2010 (van Eekelen et al. 2010). In vitro, van Eekelen et al. showed that hNSC-aaTSP-1 has an anti-angiogenic effect on human brain cells and dermal microvascular endothelial cells, when co-cultured with them established glioma cells and CD133+ glioma-initiating cells (van Eekelen et al. 2010). In vivo, a single administration of hNSC-aaTSP-1 can markedly reduce tumor vessel density, inhibit tumor progression, and increase survival of mice bearing malignant human gliomas (van Eekelen et al. 2010).

In another study, Ghaedi et al. (2011) produced human MSCs expressing alpha-1-antitrypsin (AAT) and evaluated their toxicity on human umbilical cord vein endothelial cells (HUVEC). Their results confirmed that AAT-expressing MSCs displayed higher cytotoxicity than non-transduced MSCs (Ghaedi et al. 2011). Using another anti-angiogenic factor, pigment epithelium-derived factor (PEDF), Wang et al. (2013) injected PEDF-expressing MSCs into mice bearing gliomas. The cells not only increased apoptosis of glioma cells but also prolonged mouse survival (Wang et al. 2013). In another recent study, Li et al. (2017) modified MSCs with anti-angiogenic factor, soluble fms-like tyrosine kinase-1 (sFlt-1), which can suppress angiogenesis by sequestering free vascular endothelial growth factor (VEGF) or by forming inactive heterodimers with VEGF receptor-2 (Li et al. 2017). The sFlt-1-expressing MSCs inhibited tube formation in vitro when co-cultured with HUVEC, reduced the microvessel density in mice (compared to control cells), inhibited hepatocarcinoma growth, and prolonged survival of the mice (Li et al. 2017).

2.4.6 Pro-apoptotic Proteins

Pro-apoptotic proteins belong to a group of proteins that can induce apoptosis in targeted cells. The most widely studied pro-apoptotic protein is TRAIL (tumor necrosis factor-related apoptosis-inducing ligand). TRAIL is a member of the TNF ligands which can bind to their death domain containing receptors DR4 and DR5, and induce apoptosis through activation of caspases.

MSCs were modified to express TRAIL and evaluated for treatment in various animal models of cancer. MSCs expressing TRAIL were used to treat Ewing sarcoma (EWS) (Guiho et al. 2016); these MSCs could induce apoptosis in AMG655

EWS cancer cells co-cultured, and could counteract tumor progression in orthotopic models of ESW in vivo (Guiho et al. 2016). MSCs expressing TRAIL could also induce apoptosis of melanoma cells (Jing et al. 2016), via activation of caspase-4 and caspase-8. Using the same strategy, pancreas derived MSCs expressing TRAIL showed ability to inhibit growth of pancreatic cancer cells (Moniri et al. 2012).

In a recent publication by Cafforio et al. (2017), the UC-MSCs with phenotype pIL6^{+}-TRAIL^{+}-GFP^{+} were selected and used to evaluate their effects on myeloma cells both in vitro and in vivo. These modified UC-MSCs strongly expressed TRAIL, induced apoptosis in U266 myeloma cells, and significantly reduced the tumor burden in NOD-SCID mice (Cafforio et al. 2017).

2.4.7 *Growth Factor Antagonists*

Growth factors are proteins which can trigger tumor growth. Hepatic growth factor (HGF) is the most important growth factor, affecting not only tumor growth but also metastasis. Therefore, some studies have used NK4, an HGF antagonist, to inhibit tumor growth and angiogenesis. MSCs expressing and delivering NK4 to kill cancer cells have been studied for various cancers, including liver cancer (Cai et al. 2017), gastric cancer (Zhu et al. 2014), pancreatic cancer (Sun et al. 2014b), and lung cancer (Kanehira et al. 2007). In all these studies, the effects of NK4-expressing MSCs on inhibition of cancer cell proliferation and tumor growth were very clear. In a study by Kanehira et al. (2007), the authors showed that the effects of NK4-expressing MSCs occur via inhibition of angiogenesis and lymphangiogenesis (Kanehira et al. 2007).

2.5 Clinical Applications of Engineered Mesenchymal Stem Cells for Cancer Therapy

To date, although there are many preclinical trials using engineered mesenchymal stem cells for cancer treatment, there have been a few clinical trials using MSCs to deliver various factors for anti-cancer activity (Table 2.2). The first study, reported by Carcia-Castro et al. in 2010, used MSCs to deliver oncolytic virotherapy to treat children with metastatic neuroblastoma (Garcia-Castro et al. 2010). In this study, MSCs were transduced with oncolytic adenovirus ICOVIR-5, a relatively new oncolytic adenovirus at the time. There were four children with metastatic refractory neuroblastoma in the study; the patients were autologously transplanted with MSCs infected with ICOVIR-5. The results showed that the tolerance to the treatment was excellent. There was one case of complete clinical response and three cases of complete remission 3 years after the therapy (Garcia-Castro et al. 2010).

Table 2.2 Preclinical trials for cancer using engineered mesenchymal stem cells

Cancer	Engineered mesenchymal stem cells	Effects	References
Breast cancer	Bone marrow derived MSCs expressing IFN-β	Reduced tumor growth and metastases and prolonged survival	Studeny et al. (2002)
	Bone marrow derived MSCs expressing TRAIL	Reduced tumor growth and metastases	Li et al. (2010), Reagan et al. (2012)
Lung	Bone marrow derived MSCs expressing PEDF	Reduced tumor growth and prolonged survival	Chen et al. (2012)
	Umbilical cord derived MSCs expressing TRAIL	Prolonged survival and increased tumor apoptosis	Yan et al. (2016)
Mesothelioma	Bone marrow derived MSCs expressing TRAIL	Reduced tumor growth	Sage et al. (2014)
Glioma	Adipose tissue derived MSCs expressing CDU	Tumor regression and prolonged survival	Altanerova et al. (2012)
	Adipose tissue derived MSCs expressing HSV-tK	Reduced tumor growth	Alieva et al. (2012), Amano et al. (2009)
	Umbilical cord derived MSCs expressing TRAIL	Reduced tumor growth	Kim et al. (2010), Sasportas et al. (2009)
	Bone marrow derived MSCs expressing TRAIL	Inhibits tumor growth	Nakamizo et al. (2005)
HCC	Bone marrow derived MSCs expressing Apoptin	Reduced tumor volume	Zhang et al. (2016)
	Umbilical cord derived MSCs expressing HNF4α	Reduced tumor growth	Wu et al. (2016)
	Bone marrow derived MSCs expressing IFN-β	Decreased tumor formation	Xie et al. (2013)
	Bone marrow derived MSCs expressing HSV-tK	Reduced tumor growth	Niess et al. (2011)
Pancreas	Bone marrow derived MSCs expressing HSV-tK	Reduced tumor growth and metastases	Zischek et al. (2009)
Ascites	Bone marrow derived MSCs expressing IL-12	Reduced ascites volume and prolonged survival	Han et al. (2014)
Lymphoma	Bone marrow derived MSCs expressing IL-21	Delayed tumor development and prolonged survival	Kim et al. (2015)
Prostate	Bone marrow derived MSCs expressing IFN-β	Reduced tumor weight and prolonged survival	Wang et al. (2012)

In the next study, Melen et al. (2016) investigated the effects of carrier cells on clinical outcome of children with neuroblastoma treated with high-dose oncolytic adenovirus delivered in MSCs. This study enrolled 12 patients with refractory neuroblastoma. All patients were transfused with CELYVIR (autologous human marrow derived MSCs infected with ICOVIR-5). The number of lesions pre- and post-therapy was assessed after the sixth dose by the level of serum enolase and by 123I MIBG-scintigraphy. The study showed that the tolerance to the treatment was excellent, with very mild and self-limited viral-related symptoms. From the results, the authors confirmed that multi-dosing of CELYVIR has an excellent safety profile in children with metastatic neuroblastoma (Melen et al. 2016).

2.6 Conclusion

Mesenchymal stem cells are the most common stem cells and have been clinically applied to treat many diseases. They play particular roles in the body, including migration to tumor sites. Given this unique property, scientists have used MSCs to transport various anti-tumor agents or factors to targeted tumors. To date, many preclinical trials have been performed using engineered MSCs with anti-tumor agents for cancer treatment in animals, and have shown promising results. Although there are only two clinical trials using this strategy to treat cancer in humans, these initial results illustrate the potential impact of this method to treat cancer in the near future.

Acknowledgment This study was supported by grant from the Ministry of Science and Technology via project No. DM.02.DA/15 "Assessment of the status, potential and capacity of research, application and development of stem cell technology in the field of medicine, pharmacy and agriculture."

References

Alieva M, Bago JR, Aguilar E, Soler-Botija C, Vila OF, Molet J, Gambhir SS, Rubio N, Blanco J (2012) Glioblastoma therapy with cytotoxic mesenchymal stromal cells optimized by bioluminescence imaging of tumor and therapeutic cell response. PLoS One 7:e35148

Altanerova V, Cihova M, Babic M, Rychly B, Ondicova K, Mravec B, Altaner C (2012) Human adipose tissue-derived mesenchymal stem cells expressing yeast cytosinedeaminase::uracil phosphoribosyltransferase inhibit intracerebral rat glioblastoma. Int J Cancer 130:2455–2463

Amano S, Li S, Gu C, Gao Y, Koizumi S, Yamamoto S, Terakawa S, Namba H (2009) Use of genetically engineered bone marrow-derived mesenchymal stem cells for glioma gene therapy. Int J Oncol 35:1265–1270

Batchelor TT, Sorensen AG, di Tomaso E, Zhang WT, Duda DG, Cohen KS, Kozak KR, Cahill DP, Chen PJ, Zhu M et al (2007) AZD2171, a pan-VEGF receptor tyrosine kinase inhibitor, normalizes tumor vasculature and alleviates edema in glioblastoma patients. Cancer Cell 11:83–95

Cafforio P, Viggiano L, Mannavola F, Pelle E, Caporusso C, Maiorano E, Felici C, Silvestris F (2017) pIL6-TRAIL-engineered umbilical cord mesenchymal/stromal stem cells are highly cytotoxic for myeloma cells both in vitro and in vivo. Stem Cell Res Ther 8:206

Cai C, Hou L, Zhang J, Zhao D, Wang Z, Hu H, He J, Guan W, Ma Y (2017) The inhibitory effect of mesenchymal stem cells with rAd-NK4 on liver cancer. Appl Biochem Biotechnol 183:444–459

Castleton A, Dey A, Beaton B, Patel B, Aucher A, Davis DM, Fielding AK (2014) Human mesenchymal stromal cells deliver systemic oncolytic measles virus to treat acute lymphoblastic leukemia in the presence of humoral immunity. Blood 123:1327–1335

Cavarretta IT, Altanerova V, Matuskova M, Kucerova L, Culig Z, Altaner C (2010) Adipose tissue-derived mesenchymal stem cells expressing prodrug-converting enzyme inhibit human prostate tumor growth. Mol Ther 18:223–231

Chawla-Sarkar M, Leaman DW, Borden EC (2001) Preferential induction of apoptosis by interferon (IFN)-beta compared with IFN-alpha2: correlation with TRAIL/Apo2L induction in melanoma cell lines. Clin Cancer Res 7:1821–1831

Chen X, Lin X, Zhao J, Shi W, Zhang H, Wang Y, Kan B, Du L, Wang B, Wei Y et al (2008) A tumor-selective biotherapy with prolonged impact on established metastases based on cytokine gene-engineered MSCs. Mol Ther 16:749–756

Chen Q, Cheng P, Yin T, He H, Yang L, Wei Y, Chen X (2012) Therapeutic potential of bone marrow-derived mesenchymal stem cells producing pigment epithelium-derived factor in lung carcinoma. Int J Mol Med 30:527–534

D'souza N, Burns JS, Grisendi G, Candini O, Veronesi E, Piccinno S, Horwitz EM, Paolucci P, Conte P, Dominici M (2012) MSC and tumors: homing, differentiation, and secretion influence therapeutic potential. In: Weyand B, Dominici M, Hass R, Jacobs R, Kasper C (eds) Mesenchymal stem cells-basics and clinical application II. Springer, New York, pp 209–266

Dominici M, Le Blanc K, Mueller I, Slaper-Cortenbach I, Marini F, Krause D, Deans R, Keating A, Prockop D, Horwitz E (2006) Minimal criteria for defining multipotent mesenchymal stromal cells. The International Society for Cellular Therapy position statement. Cytotherapy 8:315–317

Du J, Zhou L, Chen X, Yan S, Ke M, Lu X, Wang Z, Yu W, Xiang AP (2012) IFN-gamma-primed human bone marrow mesenchymal stem cells induce tumor cell apoptosis in vitro via tumor necrosis factor-related apoptosis-inducing ligand. Int J Biochem Cell Biol 44:1305–1314

Duan X, Guan H, Cao Y, Kleinerman ES (2009) Murine bone marrow-derived mesenchymal stem cells as vehicles for interleukin-12 gene delivery into Ewing sarcoma tumors. Cancer 115:13–22

Dwyer RM, Ryan J, Havelin RJ, Morris JC, Miller BW, Liu Z, Flavin R, O'Flatharta C, Foley MJ, Barrett HH (2011) Mesenchymal stem cell-mediated delivery of the sodium iodide symporter supports radionuclide imaging and treatment of breast cancer. Stem Cells 29:1149–1157

Francois S, Bensidhoum M, Mouiseddine M, Mazurier C, Allenet B, Semont A, Frick J, Sache A, Bouchet S, Thierry D et al (2006) Local irradiation not only induces homing of human mesenchymal stem cells at exposed sites but promotes their widespread engraftment to multiple organs: a study of their quantitative distribution after irradiation damage. Stem Cells 24:1020–1029

Friedenstein AJ, Chailakhjan RK, Lalykina KS (1970) The development of fibroblast colonies in monolayer cultures of guinea-pig bone marrow and spleen cells. Cell Tissue Kinet 3:393–403

Gao P, Ding Q, Wu Z, Jiang H, Fang Z (2010) Therapeutic potential of human mesenchymal stem cells producing IL-12 in a mouse xenograft model of renal cell carcinoma. Cancer Lett 290:157–166

Garcia-Castro J, Alemany R, Cascallo M, Martinez-Quintanilla J, Arriero Mdel M, Lassaletta A, Madero L, Ramirez M (2010) Treatment of metastatic neuroblastoma with systemic oncolytic virotherapy delivered by autologous mesenchymal stem cells: an exploratory study. Cancer Gene Ther 17:476–483

Ghaedi M, Soleimani M, Taghvaie NM, Sheikhfatollahi M, Azadmanesh K, Lotfi AS, Wu J (2011) Mesenchymal stem cells as vehicles for targeted delivery of anti angiogenic protein to solid tumors. J Gene Med 13:171–180

Guiho R, Biteau K, Grisendi G, Taurelle J, Chatelais M, Gantier M, Heymann D, Dominici M, Redini F (2016) TRAIL delivered by mesenchymal stromal/stem cells counteracts tumor development in orthotopic Ewing sarcoma models. Int J Cancer 139:2802–2811

Han J, Zhao J, Xu J, Wen Y (2014) Mesenchymal stem cells genetically modified by lentivirus-mediated interleukin-12 inhibit malignant ascites in mice. Exp Ther Med 8:1330–1334

Jing HX, Duan J, Zhou H, Hu QM, Lei TC (2016) Adipose-derived mesenchymal stem cell-facilitated TRAIL expression in melanoma treatment in vitro. Mol Med Rep 14:195–201

Johns TG, Mackay IR, Callister KA, Hertzog PJ, Devenish RJ, Linnane AW (1992) Antiproliferative potencies of interferons on melanoma cell lines and xenografts: higher efficacy of interferon beta. J Natl Cancer Inst 84:1185–1190

Kanehira M, Xin H, Hoshino K, Maemondo M, Mizuguchi H, Hayakawa T, Matsumoto K, Nakamura T, Nukiwa T, Saijo Y (2007) Targeted delivery of NK4 to multiple lung tumors by bone marrow-derived mesenchymal stem cells. Cancer Gene Ther 14:894–903

Kazimirsky G, Jiang W, Slavin S, Ziv-Av A, Brodie C (2016) Mesenchymal stem cells enhance the oncolytic effect of Newcastle disease virus in glioma cells and glioma stem cells via the secretion of TRAIL. Stem Cell Res Ther 7:149

Khakoo AY, Pati S, Anderson SA, Reid W, Elshal MF, Rovira II, Nguyen AT, Malide D, Combs CA, Hall G (2006) Human mesenchymal stem cells exert potent antitumorigenic effects in a model of Kaposi's sarcoma. J Exp Med 203:1235–1247

Kim SM, Oh JH, Park SA, Ryu CH, Lim JY, Kim DS, Chang JW, Oh W, Jeun SS (2010) Irradiation enhances the tumor tropism and therapeutic potential of tumor necrosis factor-related apoptosis-inducing ligand-secreting human umbilical cord blood-derived mesenchymal stem cells in glioma therapy. Stem Cells 28:2217–2228

Kim SW, Kim SJ, Park SH, Yang HG, Kang MC, Choi YW, Kim SM, Jeun SS, Sung YC (2013) Complete regression of metastatic renal cell carcinoma by multiple injections of engineered mesenchymal stem cells expressing dodecameric TRAIL and HSV-TK. Clin Cancer Res 19:415–427

Kim N, Nam YS, Im KI, Lim JY, Lee ES, Jeon YW, Cho SG (2015) IL-21-expressing mesenchymal stem cells prevent lethal B-cell lymphoma through efficient delivery of IL-21, which redirects the immune system to target the tumor. Stem Cells Dev 24:2808–2821

Krasikova LS, Karshieva SS, Cheglakov IB, Belyavsky AV (2015) [Mesenchymal stem cells expressing cytosine deaminase inhibit growth of murine melanoma B16F10 in vivo]. Mol Biol 49:1007–1015

Kucerova L, Matuskova M, Pastorakova A, Tyciakova S, Jakubikova J, Bohovic R, Altanerova V, Altaner C (2008) Cytosine deaminase expressing human mesenchymal stem cells mediated tumour regression in melanoma bearing mice. J Gene Med 10:1071–1082

Kucerova L, Skolekova S, Demkova L, Bohovic R, Matuskova M (2014) Long-term efficiency of mesenchymal stromal cell-mediated CD-MSC/5FC therapy in human melanoma xenograft model. Gene Ther 21:874–887

Li GC, Ye QH, Xue YH, Sun HJ, Zhou HJ, Ren N, Jia HL, Shi J, Wu JC, Dai C (2010) Human mesenchymal stem cells inhibit metastasis of a hepatocellular carcinoma model using the MHCC97-H cell line. Cancer Sci 101:2546–2553

Li X, Zhang P, Liu X, Lv P (2015) Expression of interleukin-12 by adipose-derived mesenchymal stem cells for treatment of lung adenocarcinoma. Thorac Cancer 6:80–84

Li G, Miao F, Zhu J, Chen Y (2017) Anti-angiogenesis gene therapy for hepatocellular carcinoma via systemic injection of mesenchymal stem cells engineered to secrete soluble Flt-1. Mol Med Rep 16:5799–5806

Ling X, Marini F, Konopleva M, Schober W, Shi Y, Burks J, Clise-Dwyer K, Wang R-Y, Zhang W, Yuan X et al (2010) Mesenchymal stem cells overexpressing IFN-β inhibit breast cancer growth and metastases through Stat3 signaling in a syngeneic tumor model. Cancer Microenviron 3:83–95

Liu X, Hu J, Sun S, Li F, Cao W, Wang YU, Ma Z, Yu Z (2015) Mesenchymal stem cells expressing interleukin-18 suppress breast cancer cells in vitro. Exp Ther Med 9:1192–1200

Melen GJ, Franco-Luzon L, Ruano D, Gonzalez-Murillo A, Alfranca A, Casco F, Lassaletta A, Alonso M, Madero L, Alemany R et al (2016) Influence of carrier cells on the clinical outcome of children with neuroblastoma treated with high dose of oncolytic adenovirus delivered in mesenchymal stem cells. Cancer Lett 371:161–170

Moniri MR, Sun XY, Rayat J, Dai D, Ao Z, He Z, Verchere CB, Dai LJ, Warnock GL (2012) TRAIL-engineered pancreas-derived mesenchymal stem cells: characterization and cytotoxic effects on pancreatic cancer cells. Cancer Gene Ther 19:652–658

Nakamizo A, Marini F, Amano T, Khan A, Studeny M, Gumin J, Chen J, Hentschel S, Vecil G, Dembinski J et al (2005) Human bone marrow-derived mesenchymal stem cells in the treatment of gliomas. Cancer Res 65:3307–3318

Niess H, Bao Q, Conrad C, Zischek C, Notohamiprodjo M, Schwab F, Schwarz B, Huss R, Jauch KW, Nelson PJ et al (2011) Selective targeting of genetically engineered mesenchymal stem cells to tumor stroma microenvironments using tissue-specific suicide gene expression suppresses growth of hepatocellular carcinoma. Ann Surg 254:767–774; discussion 774–765

Reagan MR, Seib FP, McMillin DW, Sage EK, Mitsiades CS, Janes SM, Ghobrial IM, Kaplan DL (2012) Stem cell implants for cancer therapy: TRAIL-expressing mesenchymal stem cells target cancer cells in situ. J Breast Cancer 15:273–282

Ren C, Kumar S, Chanda D, Kallman L, Chen J, Mountz JD, Ponnazhagan S (2008) Cancer gene therapy using mesenchymal stem cells expressing interferon-beta in a mouse prostate cancer lung metastasis model. Gene Ther 15:1446–1453

Ryu CH, Park SH, Park SA, Kim SM, Lim JY, Jeong CH, Yoon WS, Oh WI, Sung YC, Jeun SS (2011) Gene therapy of intracranial glioma using interleukin 12-secreting human umbilical cord blood-derived mesenchymal stem cells. Hum Gene Ther 22:733–743

Sage EK, Kolluri KK, McNulty K, Lourenco Sda S, Kalber TL, Ordidge KL, Davies D, Gary Lee YC, Giangreco A, Janes SM (2014) Systemic but not topical TRAIL-expressing mesenchymal stem cells reduce tumour growth in malignant mesothelioma. Thorax 69(7):638–647. https://doi.org/10.1136/thoraxjnl-2013-204110

Sasportas LS, Kasmieh R, Wakimoto H, Hingtgen S, Van De Water JA, Mohapatra G, Figueiredo JL, Martuza RL, Weissleder R, Shah K (2009) Assessment of therapeutic efficacy and fate of engineered human mesenchymal stem cells for cancer therapy. Proc Natl Acad Sci U S A 106:4822–4827

Serakinci N, Christensen R, Fahrioglu U, Sorensen FB, Dagnæs-Hansen F, Hajek M, Jensen TH, Kolvraa S, Keith NW (2011) Mesenchymal stem cells as therapeutic delivery vehicles targeting tumor stroma. Cancer Biother Radiopharm 26:767–773

Shah K (2013) Stem cell therapeutics for cancer. Wiley, Hoboken

Siegel RL, Miller KD, Jemal A (2017, Jan 1) Cancer statistics. CA Cancer J Clin 67(1):7–30.

Son BR, Marquez-Curtis LA, Kucia M, Wysoczynski M, Turner AR, Ratajczak J, Ratajczak MZ, Janowska-Wieczorek A (2006) Migration of bone marrow and cord blood mesenchymal stem cells in vitro is regulated by stromal-derived factor-1-CXCR4 and hepatocyte growth factor-c-met axes and involves matrix metalloproteinases. Stem Cells 24:1254–1264

Studeny M, Marini FC, Champlin RE, Zompetta C, Fidler IJ, Andreeff M (2002) Bone marrow-derived mesenchymal stem cells as vehicles for interferon-β delivery into tumors. Cancer Res 62:3603–3608

Sun S, Liu X, Jiang D, Lu Z, Li F (2014a) [Effect of interleukin-18 gene modified human umbilical cord mesenchymal stem cells on proliferation of breast cancer cell]. Zhonghua Yi Xue Za Zhi 94:2013–2017

Sun YP, Zhang BL, Duan JW, Wu HH, Wang BQ, Yu ZP, Yang WJ, Shan YF, Zhou MT, Zhang QY (2014b) Effect of NK4 transduction in bone marrow-derived mesenchymal stem cells on biological characteristics of pancreatic cancer cells. Int J Mol Sci 15:3729–3745

Tang Y, Li J, Zhao S, Liu J (2007) Killing effect of the herpes simplex virus thymidine kinase/ganciclovir enzyme/prodrug system on human nasopharyngeal carcinoma cells. J Int Med Res 35:433–441

van Eekelen M, Sasportas LS, Kasmieh R, Yip S, Figueiredo JL, Louis DN, Weissleder R, Shah K (2010) Human stem cells expressing novel TSP-1 variant have anti-angiogenic effect on brain tumors. Oncogene 29:3185–3195

Wang GX, Zhan YA, Hu HL, Wang Y, Fu B (2012) Mesenchymal stem cells modified to express interferon-beta inhibit the growth of prostate cancer in a mouse model. J Int Med Res 40:317–327

Wang Q, Zhang Z, Ding T, Chen Z, Zhang T (2013) Mesenchymal stem cells overexpressing PEDF decrease the angiogenesis of gliomas. Biosci Rep 33:e00019

Wu N, Zhang YL, Wang HT, Li DW, Dai HJ, Zhang QQ, Zhang J, Ma Y, Xia Q, Bian JM et al (2016) Overexpression of hepatocyte nuclear factor 4alpha in human mesenchymal stem cells suppresses hepatocellular carcinoma development through Wnt/beta-catenin signaling pathway downregulation. Cancer Biol Ther 17:558–565

Xie C, Xie DY, Lin BL, Zhang GL, Wang PP, Peng L, Gao ZL (2013) Interferon-beta gene-modified human bone marrow mesenchymal stem cells attenuate hepatocellular carcinoma through inhibiting AKT/FOXO3a pathway. Br J Cancer 109:1198–1205

Xu G, Jiang XD, Xu Y, Zhang J, Huang FH, Chen ZZ, Zhou DX, Shang JH, Zou YX, Cai YQ et al (2009) Adenoviral-mediated interleukin-18 expression in mesenchymal stem cells effectively suppresses the growth of glioma in rats. Cell Biol Int 33:466–474

Xu G, Guo Y, Seng Z, Cui G, Qu J (2015) Bone marrow-derived mesenchymal stem cells co-expressing interleukin-18 and interferon-beta exhibit potent antitumor effect against intracranial glioma in rats. Oncol Rep 34:1915–1922

Yan C, Song X, Yu W, Wei F, Li H, Lv M, Zhang X, Ren X (2016) Human umbilical cord mesenchymal stem cells delivering sTRAIL home to lung cancer mediated by MCP-1/CCR2 axis and exhibit antitumor effects. Tumour Biol 37:8425–8435

Yang X, Du J, Xu X, Xu C, Song W (2014) IFN-gamma-secreting-mesenchymal stem cells exert an antitumor effect in vivo via the TRAIL pathway. J Immunol Res 2014:318098

Zhang J, Hou L, Wu X, Zhao D, Wang Z, Hu H, Fu Y, He J (2016) Inhibitory effect of genetically engineered mesenchymal stem cells with Apoptin on hepatoma cells in vitro and in vivo. Mol Cell Biochem 416:193–203

Zhu Y, Cheng M, Yang Z, Zeng CY, Chen J, Xie Y, Luo SW, Zhang KH, Zhou SF, Lu NH (2014) Mesenchymal stem cell-based NK4 gene therapy in nude mice bearing gastric cancer xenografts. Drug Des Devel Ther 8:2449–2462

Zielske SP, Livant DL, Lawrence TS (2009) Radiation increases invasion of gene-modified mesenchymal stem cells into tumors. Int J Radiat Oncol Biol Phys 75:843–853

Zischek C, Niess H, Ischenko I, Conrad C, Huss R, Jauch K-W, Nelson PJ, Bruns C (2009) Targeting tumor stroma using engineered mesenchymal stem cells reduces the growth of pancreatic carcinoma. Ann Surg 250:747–753

Chapter 3
Stem Cells Treatment to Combat Cancer and Genetic Disease: From Stem Cell Therapy to Gene-Editing Correction

Fakher Rahim, Babak Arjmand, Bagher Larijani, and Parisa Goodarzi

The first use of umbilical cord stem cells was performed in a child with Fanconi's anemia (a type of congenital anemia), and so far, hundreds of successful stem cell transplantations have been carried out to date; thus, large storage centers have been established in different countries of the world. Researchers suggest that around 30,000 patients are diagnosed with diseases that can be treated with bone marrow stem cells, and about 75% of these patients are not able to find a suitable candidate for bone marrow donation. On the other hand, the search for bone marrow donor registry centers takes a lot of time, thus saving the cord blood to shorten the time and increasing the amount of the donor; thus, giving more time to people with genetic disorders and hematologic malignancies such as acute leukemia, anemia, and

F. Rahim (✉)
Health Research Institute, Thalassemia and Hemoglobinopathies Research Center, Ahvaz Jundishapur University of Medical Sciences, Ahvaz, Iran

B. Arjmand
Cell Therapy and Regenerative Medicine Research Center, Endocrinology and Metabolism Molecular-Cellular Sciences Institute, Tehran University of Medical Sciences, Tehran, Iran

Metabolomics and Genomics Research Center, Endocrinology and Metabolism Molecular-Cellular Sciences Institute, Tehran University of Medical Sciences, Tehran, Iran

B. Larijani
Endocrinology and Metabolism Research Center, Endocrinology and Metabolism Clinical Sciences Institute, Tehran University of Medical Sciences, Tehran, Iran
e-mail: emrc@tums.ac.ir

P. Goodarzi
Brain and Spinal Cord Injury Research Center, Tehran University of Medical Sciences, Tehran, Iran

P. V. Pham, A. El-Hashash (eds.), *Stem Cells for Cancer and Genetic Disease Treatment*, Stem Cells in Clinical Applications, https://doi.org/10.1007/978-3-319-98065-2_3

immunodeficiency that suffer from short life. Today, extensive research is devoted to the treatment of genetic and malignant diseases and lesions, repair of damaged organs and tissue, crucial organs regeneration, and repair of other damaged tissues using various type of stem cells. There are many ways to use stem cells in basic and clinical research. However, technically, there are a lot of limitations in such promising theories in treating diseases and achieving an appropriate treatment method for them, which is only possible through more research and testing. The emergence of new diseases that either did not respond to conventional treatments or did not provide therapeutic treatment at all has led researchers to use therapeutic therapies to treat more effective cell-mediated therapies for severe morbid illnesses.

3.1 Genetic Diseases

Actually, most people are familiar with the concept of genetics. For example, we say very well that the disease in the family is genetic. A lot of research has also been done that shows that most people know genetics from the gate of heredity. As much as people know genetics in terms of inheritance, traits from parents to offspring do not know genetics at the molecular level (Fareed and Afzal 2017).

3.1.1 Types of Genetic Diseases

Inside the nucleus of the human body cells and on the chromosomes there are "genes" that contain information for controlling the entire body. If each of these genes is defective and unable to transmit their information properly, the functioning of the body is disrupted and a genetic disorder is created. Genetic variations are often transmitted from parents to children and transmitted over generations. Genetic disorder is due to one or more defects in one or more genes (Song et al. 2017). This defect is in the genetic code of DNA and can disrupt the natural activity of a vital body such as the immune system or the endocrine system or prevent the growth of a normal organ such as bones or skin. Genetic diseases, due to their appearance, are divided into different categories, including single genetic disorders that are rare, and present only among 2% of the total population, and 6–8% of children hospitalized are affected by this disorder. The frequencies of this category that are prevalent in many countries especially Asian population include major thalassemia, hemophilia A and B, and mental retardation due to P.K.U and sexual obscurity caused by the adrenal gland (Shawky et al. 2012). The other category is chromosomal abnormalities that occur in 17 childbirths per 1000 live births (Yap et al. 2015).

These diseases include mental retardation such as Down syndrome and other disabilities. Overall, 50% of self-abortions in the first trimester of pregnancy are also due to a chromosomal abnormality (Dutta et al. 2011). Besides, multifactorial diseases are genetic and environmental factors involved in the onset of these diseases (Amato et al. 2017; Otsuka et al. 2017). These diseases include developmental abnormalities, congenital anomalies, and adult common diseases such as cardiovascular disease, diabetes, and MS, and one of the most common abnormalities of the fetus is the neural tube defect (Lee 2017; Momtaz et al. 2017).

In summary, it can be said that anything that causes the gene to fail properly to transmit such as chemicals, infectious diseases, x-rays, and drugs, is one of the most important causes of genetic mutations. These factors affect the genes of individuals and cause mutations, and this mutated gene is transmitted to later generations and causes disease.

3.1.2 *Causes of Genetic Diseases*

Genetic and inherited from parents, which means the child has a sperm from the beginning (such as hereditary hearing loss, thalassemia) (Rappaport 2016).

3.1.3 *Treatment of Congenital and Genetic Diseases*

Most genetic diseases are currently untreated and the only way is to prevent them from creation. The most effective way to prevent is "genetic counseling" before marriage and before pregnancy (Rappaport 2016). Some genetic diseases such as P.K.U, cleft palate, congenital dislocation of the pelvis, congenital hypothyroidism, treatment, should be diagnosed at the maximum in the first month of birth (Gamez et al. 2017; Haley 2017; Marsico and Kimberlin 2017; Skokowa et al. 2017).

3.2 Stem Cells

During the history of mankind, medical science has been associated with the great thoughts and courage of pioneers and the desire to achieve the high quality of life of those who suffer from illness is a constant concern for the human. In the past few years, medical science has undergone a major revolution and transformation that has even overshadowed developments in the discovery of penicillin, anesthetic drugs, or X-rays. The world has entered a period of bone marrow transplantation, heart, liver and kidney transplantation, or the discovery of DNA and the human genome, modern methods of laboratory infants, animal breeding, and stem cell therapy, and also very close to the production of various tissues and human organs. Stem cells are the primary cells that can transform and differentiate into different types of other cells (Tweedell 2017). Under certain physiological or laboratory conditions, these cells can be converted into cells with specific functions, such as muscle cells of the heart or insulin-producing cells in the pancreas (Ardeshiry Lajimi et al. 2013; Ebrahimi et al. 2014; Ebrahimi and Rahim 2014; Rahim et al. 2013; Saki et al. 2013; Shahrabi et al. 2014). Stem cells have two important properties that distinguish them from other cell types (Tweedell 2017). First, the regeneration ability, these cells are undifferentiated cells with the unlimited ability to reproduce. Second, pluripotent, stem cells are capable of producing any kind of cell in the body. They can be converted into cells with specific functions, such as muscle cells of the heart or insulin-producing cells in the pancreas.

3.2.1 Embryonic Stem Cells

These stem cells existed in the body of the embryo during the first weeks of its formation, which means the cells that make up the body of the human embryo (Kugler et al. 2017). It is clear that these cells can form different types of tissues and organs. These cells are taken from the internal cell mass of the 16–14-day-old fetus and are able to make all the cells and tissues of a person.

3.2.2 Cord Blood Stem Cell

The umbilical cord stem cells are from other potent cells, which, like adult stem cells, can produce a variety of cells in the laboratory. There are two types of stem cells in the umbilical cord that are able to make blood cells and bone and fat cells, as well as a replacement for bone marrow cells in bone marrow transplantation (Broxmeyer 2011; Sideri et al. 2011). At the birth of the baby, these cells can be removed by cutting the umbilical cord from the blood of umbilical veins. These cells are less capable of converting into tissues and organs than embryonic stem cells, but their differentiation is much easier. The cord matrix called Wharton's jelly is the source of adult mesenchymal stem cells.

3.2.3 Adult Stem Cells

Adult stem cells are undifferentiated cells that are found in various cells of human tissues and organs, and have the ability to regenerate and differentiate into a variety of specific cells of the body or organ (Yang et al. 2017). The initial roles of these cells in a living organism include the protection and repair of the tissues that are derived from them. Scientists have found adult stem cells in more tissues than they thought. These findings advised scientists to use these cells in transplant science.

More than 30 years now, the use of stem cells from adult hematopoietic cells that separate bone marrow from the transplant passes. Adult stem cells have been detached from the body in many organs and tissues, but the important thing is that there are very few of these cells in each tissue that reside in a particular area of that tissue for years. These hidden cells are activated with the advent of disease or tissue damage. The tissues in which adult stem cells are found are bone marrow, peripheral blood, brain, blood vessels, dental pulp, skeletal muscle, skin, liver, pancreas, cornea, retina, and digestive system.

Scientists in many laboratories are working to transform adult stem cells into cell cultures to specific types of cells for the treatment of diseases and tissue damage. The therapeutic potential of these cells is the replacement of dopamine-producing cells in the brain in Parkinson's disease, the production of insulin-like cells for the type of diabetes (insulin-dependent), and the repair of degenerated muscle cells (Ballios and van der Kooy 2010; Barkho and Zhao 2011).

3.3 Cancer Perspectives

Cancer is a type of disease in which abnormal cells become uncontrolled and can involve adjacent tissues. There are over 200 different types of cancer, each of which is created in specific ways (O'Neill et al. 2017). What is common in all of them is that they all begin in the same way, by changing the natural structure of a cell. Cancer cells can also spread through the bloodstream or lymph to other parts of the body. The human body consists of millions of cells that together create tissues such as muscles, bones, and skin, each of which has a single function. The cells replace naturally, with two folds, in a regular and controlled manner, which leads to the growth and repair of the tissues of the body (Earls et al. 2013). Sometimes the division of cells is done uncontrolled and abnormal. Abnormal and excessive growth of tissue cells is called a tumor that can be benign or malignant (Elmore 2007; Mason and Rathmell 2011). The uncontrolled division of abnormal cells, unclear when it stops, leads to the formation of a type of abnormal cell called a tumor. All tumors are not cancer. There are two types of tumors: benign and malignant (Christopher et al. 2015). Benign tumors are usually limited to a small part of the body and often grow slowly (Oldfors et al. 1988). Depending on their size and location, these tumors are often removed surgically leading to no problem to patients (Feng et al. 2017). Cancer or malignant tumors also begin in a limited space, but can spread to adjacent tissues and organs. Cancer cells may be transmitted through the bloodstream or the lymphatic system to distant areas in the body.

Types of cancer include skin-forming cancers (such as skin cancer) or lining the inside of the organs (such as lung cancer) or endocrine glands (such as breast cancer), sarcoma, cancers of tissue in the connective tissue is called cartilage, bone and muscle, lungs and lymphomas, cancers that originate from blood-forming cells and immune cells (Sudhakar 2009). In 2005, in the world, 7.5

million people died due to cancer. By 2020, 16 million people will get cancer, and at the same time, 10 million people die each year due to the disease (Roquette et al. 2017). In the next 10 years, if 85 million people die, they will die because of cancer. More than 70% of deaths from cancer are found in countries with low or moderate income (Traebert et al. 2013). But the real problem with cancer is more than these numbers. Because one-third of patients suffer from depression and anxiety at a clinical level, and because of the loss of income and the need to provide treatment costs, they are severely affected by family economic performance. However, cancer can be released, because more than 40% of cancers can be prevented and one-third of the patients can be curable if they are diagnostically timely diagnosed. In the rest of the patients that are untreatable, supportive therapies will help to improve the quality of life of patients. Cancer in developing countries is now the second leading cause of death and in developing countries is the third leading cause of death (the first cause is cardiovascular disease and the third cause is road accidents) (Khanna and Gharpure 2017). There is a significant difference in the incidence of cancer and its extent in different geographical regions of the country. Cancer research plays a central role in national cancer planning. Considering the high cost of cancer treatment at advanced stages of the disease, investing in cancer research is economically feasible. Cancer does not necessarily mean death. Fears are a natural reaction, but with better treatment, more people are saved from cancer. After treatment, cancer patients return to normal. Can cancer be triggered by blows or collisions? Sometimes you will notice a blow to your underlying skin that already existed, such as a mass in the breast. But that does not mean that it has hit the disease. Cancer is not in any way contagious in the daily routine. People do not get the disease by hanging or hugging an infected person. Only a few types of cancer can be hereditary, but it seems that in some families it is possible to have a type of cancer that is more common, such as breast, ovarian, and intestinal cancer. But in other cases, lifestyle, such as eating habits or some genetic factors, is likely to be involved, but does not mean that cancer is inherited. Diagnosis of cancer as soon as possible creates a great chance for the success of the treatment, and each cancer has its own symptoms, and if patients ignore these symptoms, cancer will spread and become more dangerous and treatment will not be easily feasible (Zhou et al. 2017). The most important symptoms you should take care of and refer to the physician for any of these are: The presence of mass anywhere in the body, such as the breast or testicles, in a blistering spot that does not heal, persistent or persistent dyspnea, indigestion, or disturbed frequent swallowing, or coughing up blood in the intestinal normal habitations, such as diarrhea or constipation, is the presence of any blood in the urine or feces and any abnormal vaginal bleeding, unjustifiably weight loss, unreasonable appetite. The exact cause is unknown, but the following factors increase the risk of cancer: smoking, diet and nutrition, obesity, infection, no physical activity, occupational exposure, genetic, alcohol, sunlight (Blackadar 2016). Cancer research has a central role in national cancer planning. Given the high cost of treating cancer in advanced stages of the disease, investing in cancer research is economically feasible.

3.4 Gene Therapy

3.4.1 Gene Therapy Perspective

Gene therapy is defined as the transfer of nucleic acids (DNA) to somatic cells of a patient which results in a therapeutic effect, and is a therapeutic approach currently in the research stage and at this stage there are various ways to modify the gene (Bera and Sen 2017; Rouanet et al. 2017; Witt et al. 2017). The most important method is to place a healthy gene in the genome of the patient. Gene therapy is currently in the research stage, and no gene therapy products approved by the US Food and Drug Administration (FDA), as well as clinical trials have also reported little success (Qazi and Hamrah 2013; Schakman and Thissen 2006; Yoshimura et al. 2010). The first gene therapy was done in 1990, which was performed on a 4-year-old child with a rare severe genetic disorder (SCID). Due to the lack of normal immune system, this patient should be kept in sterile conditions and taking regular high doses of antibiotics. In this patient, the white blood cells were taken and then they were grown in the laboratory and the healthy gene was introduced into the bloodstream and returned to the patient again. Further studies showed that the patient's immune system was boosted and the patient returned to normal life, but this treatment was not a lifelong treatment, and the patient had to repeat it every few months (Wirth 2011). The fact is that because of the overwhelming biological complexity of gene therapy in humans and the lack of complete recognition of many genetic diseases, the use of gene therapy in humans is not possible (Mac Gabhann et al. 2010; Naylor and Chen 2010). Another critical issue is the medical ethics, and it should be noted that gene therapy on somatic cells (non-sexual) and sex cells is different (Farajkhoda 2017; Sheppard et al. 2016). If gene therapy is performed on germs, it can also be transmitted to subsequent generations, while in gene therapy on somatic cells only the recipient gene is improved and the gene is intended to be inherited by subsequent generations.

Some people protest against any genetic changes in humans without considering their benefits. Many others only confirm gene therapy in non-sexual cells (somatic cells), but do not accept the use of gene therapy in the sex cells. However, there are also some who believe that gene therapy in sex cells can also be introduced with new laws. The possibility of treating a non-curable disease by a gene therapy method has been taken into consideration, and the following are in the confirmation of gene therapy, include gene therapy provides the complete treatment of the disease and does not only eliminate the symptoms, in sexual cells may be the only way to treat some genetic diseases., and by preventing a defective gene transfer to subsequent generations, treatment of that generations will not be necessary because the illness is not inherited from the parents treated with gene therapy. Some people also emphasize that it is very difficult to engage patients in long-term clinical research. A patient undergoing gene therapy should be monitored for decades to be informed of the effects of gene therapy in later generations. Some also believe that children are not able to decide on gene therapy because of their age, and they do not properly understand their benefits and prospects. Briefly, there are a number of issues involved in the rejection of gene therapy in sex cells, including (1) lack of sufficient scientific information and high clinical risk, as well as the long-term effects of this type of treatment; (2) such experiments may ultimately lead to encouraging other researchers to work on genetic changes in people who are healthy; (3) because gender biology in fetuses requires the use of fetuses and, as a result, is important for the birth of a baby, such research will lead to the development of research projects that a person (a fetus after birth) has been tested without prior consent; (4) the high cost of gene therapy can in summary be seen in the near future in the treatment of a number of genetic diseases or cancers in the gene therapy method, which will definitely revolutionize medical science and will open the way for the treatment of many diseases.

Gene therapy is a way to correct defective genes that cause the disease. There are various ways to correct a defective gene, the most important of which is the insertion of a healthy gene in the patient's genome. In most gene therapy studies, a healthy gland is placed instead of defective in order to modify the gene that produces the disease, and a gene-bearing molecule is called the vector, which is used to transfer the gene to the patient's cell. One of the most important vectors used is genetically engineered viruses that can carry human DNA (Bailer et al. 2017; Lau and Hemsley 2017; Naso et al. 2017). Viruses have the ability to transfer their genes to human cells, resulting in the disease, which researchers use this property and transfer the genes to human cells with viruses. Target cells in the patient are infected by genetically modified viruses that result in a healthy human germ in the cell, which results in the production of a protein that the patient has not previously been able to produce. As a result, in addition to viruses, other non-viral vectors are also used. Liposomes are another way of transferring genes into cells (Mohammadian et al. 2017; Riley and Vermerris 2017; Saffari et al. 2016). Liposomes are spheres in which the center of the target gene and surrounding fat layer are located, thus allowing them to cross the cell membrane and thereby liberate the gene within the cell. Another method is the use of polycatyonal complexes capable of binding to the gene (Murai 2017). The polymer and gene blend are capable of phagocytosis by the cell, resulting in the gene being introduced into the cell and then expressed. The research on the last two methods has recently been considered more widespread, the main reasons for which are the disadvantages of using viruses as vector (Jayant et al. 2016; Warnock et al. 2011). One of the other important reasons that why research into the use of liposomal vectors has led to economic challenges, is the final product (Fougeroux and Holst 2017). In the two methods mentioned above, gene therapy is temporary and the patient has to be treated at intervals, which results in more economic attractiveness for pharmaceutical companies. What researchers did in this field is to attach *policosanol* vectors to complexes, then adjoin the DNA to them; because it has a negative charge DNA, it binds to a polycation that has a positive charge. Next, they place it next to human cells or animals in the laboratory, which is called the phagocyte linking of the gene to the polymer. Thus, the complex can be reached, in the cell, it reaches the local nucleus where the gene is based, and expressing it is to express, namely, the work to be done, to start. Ultimately, they want to make a combination that can transfer the desired gene to the patient by mixing the gene with a polymer and producing a medicinal product that the gene is intended to enter into patient tissue and the disease is resolved.

3.4.2 *The Future of Human Gene Therapy*

Over the past decades, noteworthy advances were done in a number of enabling technologies, in the molecular understanding of diseases and the development of vectors. Gene therapy is a complex process, involving multiple steps in the human body, including delivery to organs, tissue targeting, cellular trafficking, regulation of gene expression level and duration, biological activity of therapeutic protein, safety of the vector and gene product, to name just a few. Most of these steps are not entirely known. The essential of effective gene therapy include therapeutically appropriate genes, suitable gene delivery systems, proof of principle of efficacy and safety in proper preclinical models, and right developed and analytical processes to provide well-defined gene therapy products for clinical studies. Hemophilia is the most promising areas for gene therapy nowadays, as well as cardiovascular diseases. For other diseases such as cancer, additional progresses in gene delivery vectors and gene expression systems will be required. The technical improvements together with the ever growing knowledge and practice in the field will certainly lead to the understanding of the full potential of gene therapy in the future (Rubanyi 2001).

3.5 Stem Cell Gene Therapy

Researchers have introduced stem cell gene therapy as a new method that can ultimately be used to treat some diseases such as Duchene's muscular dystrophy (Ikezawa et al. 2003; Morgan and Partridge 1992; Tedesco et al. 2012). They reported that stem cell gene therapy is applicable to 60% of patients with dystrophy. In the United States, 1 in 5000 boys suffer from Duchene's muscular dystrophy, so the disease is considered as the most commonly reported lethal genetic disorder. In this study, researchers took skin cells and inject to patients and reprogrammed the cells into induced pluripotent stem cells. Then, using the "CRISPR/Cas9" technology, the modified dystrophin gene was removed. The researchers then found that these iPS cells lack dystrophin mutations that differentiate the pathways of the muscle cells and linked these cells to mice that had dystrophin mutations. They found that these transplanted muscle cells can successfully produce dystrophin protein. This is the first study to revive humanized iPS cells that have been able to directly revitalize the function of dystrophy muscle tissue (Liu et al. 2007; Tedesco et al. 2012). Moreover, stem cell gene therapy has assisted ophthalmologic surgeons to create new era in the retinal surgery. Nowadays there are good approaches to heal damaged eyes, besides today stem cell gene therapy has found new methods for treating retinal injuries (Ochakovski et al. 2017); thus, this type of treatment can be used for genetic and congenital diseases, so that the patient may get perfect vision (Cuenca et al. 2014).

3.6 Perspectives on Stem Cell Gene Therapy

3.6.1 Targeting Stem Cell Gene Therapy to Genetic Diseases

Genetic and rare genetic diseases in the country are mostly due to familial marriages and said that the treatment of these diseases is based on gene editing or gene therapy that is a branch of regenerative medicine, in which the cells are engineered to modify the patient's cells. So far many efforts have done to treat genetic and rare diseases with gene modification globally (Table 3.1). Nowadays, gene transfer into hematopoietic stem cells (HSCs) considers as a form of established therapy for various genetic diseases (Galeev et al. 2017; Masiuk et al. 2017; Zhang 2017). Ideal pathological target for HSC gene therapy includes primary immunodeficiency disease (PID), adenosine-deaminase (ADA) deficiency, other diseases including X-linked severe combined immunodeficiency disease (X-SCID), chronic granulomatous disease (CGD) and Wiskott–Aldrich syndrome (WAS) (Emery 2004; Williams 2014). Not only the efficacy of this therapy considerably varies, but accompanies genomic insertion most of the patients may subject to an inherent risk of insertional leukemogenesis (Otsu 2015); thus, researchers focus on continuing to further improve and take full advantage of both safety and efficacy of HSC gene therapy (Kaufmann et al. 2013). Patient- or disease-specific iPSCs as a platform for modeling genetic disorders has come to reality, recently (Avior et al. 2016). These model cells provide a specific genetic disease modeling platform for preclinical studies, as well as to use for testing gene-modification procedure safety and efficacy.

3.6.2 Targeting Stem Cell Gene Therapy to Cancer

In recent years, the advancement of medical knowledge in the field of cancer has progressed dramatically in recent years. One of these treatments is the transplantation of the brain and bone-derived stem cells to treat some of the diseases of the blood, cancer, genetics, and immune disorders. Recently, targeting stem cell gene therapy to cancer by many researchers (Mavroudi et al. 2014) has led to get aware with the use of stem cells in the treatment of malignant diseases (Table 3.2). Stem cells are used to treat malignant diseases such as various cancers, especially in cancer like pancreas or advanced gastric cancer of which routine treatments cannot do much (Yoshida et al. 2004). Current treatment of prostate cancer, especially hormone-resistant cancers, may result in severe consequences due to side effects

Table 3.1 Available studies on stem cell gene therapy in various genetic diseases

Authors	Country	Method Stem cell	Disorder	Target gene	Findings
Niu et al. (2016)	China	CRISPR/Cas9 iPSCs	β-Thalassemia (β-Thal)	Homo sapiens hemoglobin β (HBB)	Provides an efficient and safe strategy to correct the HBB gene mutation
Firth et al. (2015)	USA	CRISPR/Cas9 iPSCs	Cystic fibrosis (CF)	Cystic fibrosis transmembrane regulator (CFTR)	This system could be adapted for the development of new therapeutic approaches
Flynn et al. (2015)	UK	CRISPR/Cas9 iPSCs	Chronic granulomatous disease (CGD)	Cytochrome b-245 heavy chain (CYBB)	Provides proof of principle for a gene therapy approach to CGD treatment using CRISPR-Cas9
Menon et al. (2015)	USA	iPSCs	X-linked severe combined immunodeficiency (SCID-X1)	Interleukin-2 receptor gamma chain (IL-2Rγ)	Study highlights the potential for the development of autologous cell therapy
Wu et al. (2015)	China	CRISPR/Cas9 SSCs	–	Crystallin, gamma C (*CRYGC*)	Efficient gene editing and provide the proof of principle of curing a genetic disease via gene correction
Navarro et al. (2014)	Spain	iPSCs derived hematopoietic cells	Fanconi anemia (FA)	Breast cancer 2 protein (*BRCA2*)	Combined procedures of gene therapy and cell reprogramming will require further improvements to enhance the genetic stability
Sun and Zhao (2014)	USA	iPSCs TALENs	Sickle cell disease (SCD)	Homo sapiens hemoglobin β (HBB)	An important first step of using cell and gene therapies for the treatment of genetic diseases such as SCD
Wang et al. (2012)	China	iPSCs	β-thalassemia major (Cooley's anemia)	β-globin	Provides strong evidence that iPSCs generated from a patient with a genetic disease have potential clinical uses
Deyle et al. (2012)	USA	iPSCs iMSCs	Osteogenesis imperfecta (OI)	Type I collagen genes	Combined gene targeting and iPSC derivation can be used to produce potentially therapeutic cells from patients with genetic disease

(continued)

Table 3.1 (continued)

Authors	Country	Method Stem cell	Disorder	Target gene	Findings
Ramnaraine et al. (2012)	USA	OcPCs	Bone cancer	Marker gene (GFP)	Study represents a step towards considering human gene therapy for bone cancer
Zhang et al. (2011)	China	iPSCs	Wilson's disease	ATP7B gene	iPSCs could be an option for autologous transplantation in Wilson's disease
Porada and Almeida-Porada (2010)	USA	MSCs	–	–	Mesenchymal stem cells as immunomodulatory/anti-inflammatory agents destroy malignant cells
Bauer Jr. et al. (2008)	USA	Foamy virus vectors/ HSc	Canine leukocyte adhesion deficiency (CLAD)	–	These results represent the first successful use of a foamy virus vector to treat a genetic disease
Sampaolesi et al. (2006)	Italy	Mesoangioblasts (vessel-associated stem cells)	Duchenne muscular dystrophy	Intron 6	Mesoangioblasts as candidates for future stem cell therapy for Duchenne patients
Gammaitoni et al. (2006))	Italy	HSCs	Nonobese diabetic/severe	–	Retroviral gene transfer into HSCs, followed by ex vivo expansion, could facilitate long-term engraftment of genetically modified HSCs
Bigger et al. (2006)	UK	HSc	Hemophilia B	Factor IX	The potential of a stem cell approach to gene delivery to tolerize recipients to a secreted foreign transgenic protein
Williams et al. (2005)	USA	Fatal bone marrow failure	Fanconi anemia (FA)	–	A gene therapy approach for treating this disease, including phase I trials
Pawliuk et al. (1999)	USA	HSc	Erythropoietic protoporphyria	–	Green fluorescent protein results in complete and long-term correction of skin photosensitivity in all transplanted mice

iPSCs induced pluripotent stem cells, *SSCs* spermatogonial stem cells, *TALENs* transcription activator-like effector nucleases, *iMSCs* induced mesenchymal stem cells, *OcPCs* osteoclast precursor cells, *HSC* hematopoietic stem cells

Table 3.2 Available studies on stem cell gene therapy in various cancers

Authors	Country	Method Stem cell	Disorder	Target gene	Findings
Wang et al. (2017)	China	Adipose-derived stem cells	Breast cancer	(CNGRCLLII (KLAKLAK)2) named CNAK	New peptide CNAK containing Asn-Gly-Arg motif had anti-angiogenic activity in vitro and in vivo
Long et al. (2017)	China	–	CD133-positive (CD133+) ovarian cancer cells	CD133	Cre/LoxP system can effectively eliminate CD133+ ovarian cancer stem cells, representing a novel therapeutic strategy for the treatment of ovarian cancer
Gao et al. (2016)	USA	iPS	Myotonic dystrophy type 1 (DM1)	polyA signals (PASs)	iPS cells will have broad clinical application in developing autologous stem cell therapy for DM1
Zhang et al. (2016)	China	MSCs	Hepatocellular carcinoma (HCC)	–	Resolve immune rejection induced by repeated applications of drug directly delivered by Ad vectors and reduce the high cost of a large-scale production and purification of exogenous drugs
Kaneti et al. (2016)	Israel	Nanoghosts derived from MSC	Multiple cancers	–	For a cancer-toxic gene inhibited the growth of metastatic orthotopic lung cancer and subcutaneous prostate cancer models and dramatically prolonged the animals' survival
Yin et al. (2016)	USA	AD-MSCs	Ovarian cancer	–	As a result, induced significant ovarian cancer cell death in vitro and in vivo
Li et al. (2016)	China	hAFS	Lung carcinoma	CXCR4	hAFS cells are capable of carrying Ad-CXCR4-DAL-1 vectors, specifically targeting to lung cancer, and causing oncolytic effects when administered in vivo
Lakota et al. (2015)	Slovakia	MSCs	Metastatic head and neck cancer	–	We did not observe any therapeutic effect. However, a temporal bicytopenia was observed
Krasikova et al. (2015)	Russia	MSCs	Melanoma	–	MSCs encoding CDA/UPRT suicide gene have a significant antitumor effect
Alexiades et al. (2015)	USA	NSC	Brain tumor	VEGFA	NSC carriers and optimizing the delivery of therapeutic payloads to disseminated tumor burdens

(continued)

Table 3.2 (continied)

Authors	Country	Method Stem cell	Disorder	Target gene	Findings
Leten et al. (2015)	Belgium	BM-MAPCs	Glioblastoma	–	Stem cell-based therapeutic approaches for glioblastoma and hereby contributes to a better understanding and optimization
Xu et al. (2015)	China	BMSCs	Glioma	Interleukin-18 and interferon-β	Delivery of IL-18 and IFN-β by BMSCs may be an excellent and promising approach to develop an effective treatment protocol for glioma therapy
You et al. (2015)	China	AF-MSCs	Ovarian tumor	Protein-human interleukin-2 (pEGFP-hIL-2)	AF-MSCs may serve as transporters for therapeutic genes targeting ovarian tumor site
Hoyos et al. (2015)	USA	MSCs/Caspase 9 (Ad.iC9)	Lung Cancer	iC9	Use of this innovative approach for the treatment of NSCLC
Iizuka et al. (2015)	Japan	iPSCs	Myeloid leukemia	RUNX1	iPSCs generated from patient-derived cells could provide a promising clinical application for treatment of the disease
Choi et al. (2015)	Republic of Korea	hAT-MSC	Brainstem glioma	sTRAIL	Gene therapy for brainstem gliomas in terms of therapeutic effects and safety
Yi et al. (2015)	Republic of Korea	NDEPT	Endometrial cancer	IFN-β	Therapeutic stem cell-based immune therapy involving the targeted expression of CD and IFN-β genes at endometrial cancer sites
Matuskova et al. (2015)	Slovakia	AT-MSC	Lung metastases	–	Combined gene-directed enzyme/prodrug therapy mediated by MSC exerted synergic cytotoxic effect
Knoop et al. (2015)	Germany	MSCs	Metastatic colon cancer	RANTES/CCL5	This novel gene therapy approach opens the prospect of NIS-mediated radionuclide therapy of metastatic cancer after MSC-mediated gene delivery
Tyciakova et al. (2015)	Slovakia	MSCs	Melanoma xenograft	Tumor necrosis factor alpha (TNFα)	Tumor cells respond to hTNFα-based treatment mediated by genetically engineered AT-MSC/hTNFα both in vitro and in vivo

Jeong et al. (2015)	Republic of Korea	MSCs	Metastatic hepatoma	MCP-1/CCL2	Treatment with irradiation and MSCs/IL-12 may be a useful strategy for enhancing antitumor activity in metastatic hepatoma
Bahrambeigi et al. (2015)	Iran	ADSCs	Melanoma	Interleukin-2 (IL2)	Therapies utilizing IL2 have to be taken into careful consideration
Zhu et al. (2014b)	China	MSCs	Gastric cancer xenografts	NK4	Studies are warranted to explore the efficacy and safety of the MSC-based NK4 gene therapy in animals and cancer patients
Connelly et al. (2014)	USA	iPSCs	Familial platelet disorder/acute myeloid leukemia (FPD/AML)	RUNX1	iPSCs confirm that RUNX1 mutations are responsible for megakaryopoietic defects in FPD patients
Abrate et al. (2014)	Italy	MSC	In prostate cancer	–	Therapeutic potential of specifically engineered MSC in prostate cancer disease
Kucerova et al. (2014)	Slovakia	MSC	Melanoma xenograft	EGFP-A375	The CD-MSC/5FC regimen can be regarded as a very effective antitumor approach to achieve long-term tumor-free survival
Kim et al. (2014b)	Korea	MSC-TRAIL	Malignant glioma	–	TMZ and MSC-TRAIL is a potential therapeutic strategy for improving the treatment of malignant gliomas
Zhang et al. (2014a)	China	PMSCs	Colorectal cancer	–	Potential of PMSCs-based gene therapy for the targeted delivery of therapeutic proteins in cancer
Purwanti et al. (2014)	China	iPS cell-derived EPCs	Metastatic breast cancer	–	iPS cell-derived EPCs have the potential to serve as tumor-targeted cellular vehicles for anticancer gene therapy
Teng et al. (2014)	USA	NSCs	Glioblastoma	–	Potential of systemic injection of NSCs to deliver anticancer agents, such as TRAIL, which yields glioma regression when combined with Lan C
Zhu et al. (2014a)	Singapore	NSCs	Breast cancer	CD40L	Particularly valuable for metastatic cancer treatments after systemic stem cell administration
Zhang et al. (2014b)	China	BMSCs	Pulmonary metastasis	B16F10	A metastatic melanoma model and cytomegalovirus-thymidine kinase expressing stem cells

(continued)

Table 3.2 (continied)

Authors	Country	Method Stem cell	Disorder	Target gene	Findings
Lopez-Ornelas et al. (2014)	México	NSCs	Glioblastoma multiforme (GBM)	Growth arrest specific-1 (Gas1)	tGas1 has an antineoplastic effect, and the results support the potential of tGas1 as an adjuvant for the treatment of gliomas
Deng et al. (2014)	China	BM-MSCs	Liver cancer	–	BM-MSC can provide a steady source of stTRAIL and may be suitable for use in the prevention of the recurrence of hepatocellular carcinoma
Zhang et al. (2014c)	China	hUCMSCs	Ovarian cancer	Interleukin-21 (IL-21)	hUCMSCs did not form gross or histological teratomas up to 60 days posttransplantation in murine lung, liver, stomach, and spleen
Ahn et al. (2013)	Republic of Korea	AT-MSCs	Canine melanoma	IFN-β	May prove useful in further explorations of the application of these combined approaches to the treatment of malignant melanoma and other tumors
Kim et al. (2013a)	Republic of Korea	GESTECs	HeLa cervical cancer cells	IFN-β	GESTECs transduced with CD gene and IFN-β may provide a potential of a novel gene therapy for anticervical cancer treatments
Yi et al. (2013)	Republic of Korea	GESTECs	Colorectal cancer	IFN-β	GESTECs is a novel strategy where the combination of the migration capacity of stem cells as a vector for therapeutic genes towards colorectal cancer
Dembinski et al. (2013)	USA	MSCs	Ovarian cancer	IFN-β	Ovarian carcinoma engrafts MSCs to participate in myofibrovascular networks and that IFN-β produced by MSCs intratumorally modulates tumor kinetics
Kim et al. (2013b)	Republic of Korea	MSC/dTRAIL-TK	Metastatic renal cell carcinoma	–	Therapeutic approach to completely eradicate metastatic tumors by simple, repeated administrations of MSC/dTRAIL-TK
Jin et al. (2013)	China	NSCs	Glioma	–	Engineered NSCs and providing a potential novel strategy for enhancing the effectiveness of gene therapy in other diseases

Yan et al. (2013)	China	MSCs	Non-Hodgkin's lymphoma (NHL)	scFvCD20-sTRAIL	scFvCD20-sTRAIL secreting HUMSCs is a novel and efficient therapeutic approach for the treatment of non-Hodgkin's lymphoma
Lee et al. (2013)	Republic of Korea	MSC/TRAIL	Malignant fibrous histiocytoma	–	MSC-TRAIL could be extended to other sarcomas and recurrent metastatic cancers for cell-mediated cancer therapy
Zhu et al. (2013)	China	MSCs	Gastric cancer	LIGHT (TNFSF14)	UCBMSCs carrying the LIGHT gene have the potential to be used as effective delivery vehicles in the treatment of gastric cancers
Ma et al. (2012)	China	UCBMSCs	Stomach cancer	LIGHT	Transgenic UCBMSCs secrete LIGHT in a paracrine manner, which has inhibition and killing effects on stomach carcinoma
Kim et al. (2012c)	Republic of Korea	MSC-TRAIL	Malignant glioma	–	MK886 combined with MSC-based TRAIL gene delivery may represent a novel strategy for improving the treatment of malignant gliomas
Zhang et al. (2012)	China	TRAIL-MSCs	Hepatocellular carcinoma	–	The potential of combination of chemotherapeutic agents and MSCs as a gene vector in the therapy of HCC
Moniri et al. (2012)	Canada	MSCs	Pancreatic cancer	PANC1, HP62, ASPC1, TRM6, and BXPC3	MSCs exhibit intrinsic inhibition of pancreatic cancer cells and that this effect can be potentiated by TRAIL-transfection on death receptor-bearing cell types
Kosaka et al. (2012)	Japan	MSCs	Glioma	AdexCAEGFP	Gene therapy and MSCs as a targeting vehicle represents a potential new therapeutic approach for malignant glioma
Mao et al. (2012)	China	MSCs	Gastric cancer	TNF-α	MSC-TNF-α can obviously inhibit gastric cancers growth in nude mice, indicating that UCBMSCs may have the potential to become a prevention approach against gastric cancer
Ryu et al. (2012)	Republic of Korea	MSCs-TK	Glioma	Cytosine deaminase (CD)	Rational for designing novel experimental protocols to increase bystander killing effect against intracranial gliomas using MSCs-TK and VPA

(continued)

Table 3.2 (continied)

Authors	Country	Method Stem cell	Disorder	Target gene	Findings
Kim et al. (2012a))	Republic of Korea	NSCs	Glioma	Cytosine deaminase (CD)	Engineered NSCs encoding suicide gene CD could provide clinical application of suicide gene therapy for patients with glioma
Fei et al. (2012)	China	MSC-CD/eGFP	Glioma	Cytosine deaminase (CD)	MSC-CD/eGFP-mediated gene therapy promoted tumor cell apoptosis in rat C6 gliomas
Zolochevska et al. (2012)	USA	ASC	Prostate cancer	Melanoma differentiation associated gene-7 (MDA-7)	Suggesting ASC-cytokine therapies might have translational applications, especially the PEDF modality
Choi et al. (2012)	Republic of Korea	hMSCs	Brainstem gliomas	–	hAT-MSCs can be easily prepared and genetically modified as cellular vehicles for prodrug gene therapy
Zhao et al. (2012)	Singapore	NSCs	Glioma	–	Production of cellular vehicles suitable for targeted anticancer gene therapy
Bexell et al. (2012)	USA	MSCs	Glioma	D74(RG2), N32, and N29 gliomas	MSCs as vectors in glioma gene therapy, although intratumoral MSC implantation provides a dense and tumor-specific vector distribution
Warrick et al. (2012)	France	Human skin stem cell	Xeroderma pigmentosum (XP)	–	Long-term genetic correction of XP-C epidermal stem cells constitutes the first preclinical model of ex vivo gene therapy for XP-C patients
Hu et al. (2011)	China	UBMC-MSCs	Ovarian cancer	Interleukin-21 (IL-21)	UBMC-MSCs can serve as vehicles for IL-21 gene delivery and inhibit the established tumor
Kim et al. (2012b)	Republic of Korea	GESTECs	Gastric cancer	Interferon-β (IFN-β)	GESTECs are a potential anticancer therapy candidate that may result in minimal side effects compared to the conventional chemotherapy
Niess et al. (2011)	Germany	MSCs	Hepatocellular carcinoma (HCC)	Herpes simplex virus thymidine kinase (HSV-Tk)	Exogenously added MSCs are recruited to growing HCC xenografts with concomitant activation of the CCL5 or Tie2 promoters within the MSC

Amano et al. (2011)	Japan	MSCs	Glioma	HSVtk gene (MSCtk)	When using a slowly growing vector cell as MSCtk, GCV should be administered soon after MSCtk implantation
Seol et al. (2011)	Republic of Korea	NSCs	Brain metastasis from breast cancer	F3.CE and CPT-11	This strategy may be considered as an effective treatment regimen for brain metastases
Seo et al. (2011)	Republic of Korea	AT-MSC	Melanoma	Interferon-β	AT-MSC-mediated IFN-β in treating canine and human cancer patients
Yi et al. (2011)	Republic of Korea	GESTECs	Lung cancer	Interferon-β (IFN-β)	GESTECs expressing fusion genes (i.e., CD and IFN-β) may have a synergic antitumor effect on lung cancer cells
Ahmed et al. (2011)	USA	NSCs	Glioblastoma	CRAD-S-pk7	NSC-based carriers have the potential to improve the clinical efficacy of antiglioma virotherapy by not only protecting therapeutic virus
Dwyer et al. (2011)	Ireland	MSC-NIS	Breast cancer	–	MSC migration and transgene expression noninvasively in real time before therapy is a major advantage to this strategy
Zhao et al. (2011)	China	UC-MSCs	Ovarian carcinoma	IL-12	Human UC-MSCs with IL-12 gene transduction can inhibit the proliferation and induce apoptosis of ovarian carcinoma
Kim et al. (2010)	Canada	GESTECs	Ovarian carcinoma	–	GESTECs expressing CD/CE may have a potent advantage to selectively treat ovarian cancers
Chang et al. (2010)	Korea	MSCs	Brain tumor	–	Allogeneic stem cells for the treatment of recurrent glioblastomas through repetitive treatments
Gao et al. (2010b)	China	MSCs	Hepatocellular carcinoma in nude	–	Potential of hMSCs as an effective delivery vehicle for therapeutic genes in the treatment of HCC
Jiang et al. (2010)	China	MSC-EN cells	Ovarian cancer	–	MSCs could be utilized as a powerful delivery system of endostatin. The endostatin produced by MSC-EN cells could inhibit the proliferation of SKOV3 cells
Amano et al. (2009)	Japan	MSCtk cells	Glioma	–	MSCtk and ganciclovir (MSCtk therapy) is more feasible and practical for clinical application than the method using neural stem cells

(continued)

Table 3.2 (continied)

Authors	Country	Method Stem cell	Disorder	Target gene	Findings
Zischek et al. (2009)	Germany	MSC	Pancreatic carcinoma	–	The active homing of MSCs into primary pancreatic tumor stroma and activation of the CCL5 promoter was verified using eGFP- and RFP-reporter genes
Gao et al. (2010a)	China	MSCs	Renal cell carcinoma	IL-12 protein	Adult MSC constitutively producing IL-12 to reduce the growth of RCC and enhance the tumor-bearing mouse survival
Joo et al. (2009)	Korea	NSCs	Breast cancer brain metastases	–	Combined with systemic administration of the prodrug 5-FC, is an effective treatment regimen for brain metastases of tumors
Bexell et al. (2009)	Sweden	MSCs	Gliomas	–	MSCs might represent a particularly well-suited vector system for delivering molecules to affect tumor angiogenesis
Ren et al. (2008)	USA	MSCs	Prostate cancer lung metastasis	Interferon (IFN)-beta	Potential of MSC-based IFN-beta therapy for prostate cancer lung metastasis
Kanehira et al. (2007)	Japan	NK4-MSCs	Multiple lung tumors	Hepatocyte growth factor (HGF)	MSCs can serve as a vehicle of gene therapy for targeting multiple lung metastatic tumors
Zhong et al. (2007)	China	HSc	Hepatocellular carcinoma	–	Liver stem cells are a potential tool for a targeted gene therapy of HCC
Hakkarainen et al. (2007)	Finland	MSCs	Lung and breast tumors	–	MSCs loaded with oncolytic adenoviruses might be a useful approach for improving the bioavailability of systemically administered oncolytic adenoviruses
Rachakatla et al. (2007)	USA	UCMS	Lung tumor	UCMS-IFN-beta	UCMS cells can potentially be used for targeted delivery of cancer therapeutics
Xin et al. (2007)	Japan	MSCs	Lung tumor	–	MSCs can be used as a vehicle for introducing biological agents into multiple lung tumor tissues

Siapati et al. (2007)	UK	HSCs	Murine leukemia	–	Tumors arising in these type of gene therapy protocols are not necessarily due to vector insertional mutagenesis
Stoff-Khalili et al. (2007)	USA	hMSCs	Lung metastases of breast carcinoma	–	hMSCs may be an effective platform for the targeted delivery of CRAds to distant cancer sites such as metastatic breast cancer
Kim et al. (2006)	USA	NSC	Medulloblastoma	–	NSCs as an effective delivery system to target and disseminate therapeutic agents to medulloblastoma
Li et al. (2006)	China	BMSCs	Chronic myelogenous leukemia (CML)	hIFN-gamma	Important potential use in the clinical gene therapy of CML
Uzzaman et al. (2005)	USA	Embryonic stem cell (ESC)	Brain tumor	–	ESC-derived astrocytes can be used as delivery vectors for CNS tumors
Zhong et al. (2005)	China	Hepatocellular carcinoma (HCC)	Hepatocellular carcinoma	–	Using migratory liver stem cells as a delivery vehicle for gene therapy for HCC
Deng et al. (2005)	China	Umbilical blood stem cells	Melanoma	Human ribonuclease inhibitor (hRI)	Capacity of RI to inhibit the tumor growth by blocking the vasculature in tumor
Nakamura et al. (2004)	Japan	MSCs	Glioma	Interleukin-2 (IL-2)	Gene therapy employing MSCs as a targeting vehicle would be promising as a new therapeutic approach for refractory brain tumor

iPS induced pluripotent stem, *MSCs* mesenchymal stem cells, *AD-MSCs* adipose-derived mesenchymal stem cells, *NSC* neural stem cell, *BM-MAPCs* bone marrow multipotent adult progenitor cells, *hAFS* human amniotic fluid stem cells, *AF-MSCs* amniotic fluid mesenchymal stem cells, *hAT-MSC* human adipose-derived mesenchymal stem cells, *NDEPT* neural stem cell-directed enzyme/prodrug therapy, *hUCMSCs* human umbilical cord mesenchymal stem cells, *GESTECs* genetically engineered stem cells, *UCBMSCs* umbilical cord blood mesenchymal stem cells, *ASC* adipose-derived stromal/mesenchymal stem cells, *HSc* hepatic stem cells caused by cytotoxic agents. Self-destructive gene therapy has been investigated for its ability to target cancer cells as an alternative to current chemotherapy. However, classical self-destructive gene therapy has several severe side effects, including the risk of immune system due to viral vector. Recently, stem cells have been considered as a new vector cell due to their ability to be linked. Suicidal gene therapy using a genetically engineered stem cell has the advantage of being safe because injection of this precursor not only destroys cancer cells, but also resilient stem cells. The attractiveness of cancer gene therapy by targeted tumor stem cells is based on activating precursors directly within the tumor mass, thus preventing systemic toxicity. The therapeutic benefits alongside proven mechanisms and limitations of using stem cells in prostate cancer have been reported recently (Kim et al. 2014a). Gastric cancer is the third commonest cause of cancer mortality globally, thus there is an urgent need for novel treatment modalities. Researcher investigated MSC-delivered NK4 gene therapy in animal model of gastric cancer, and reported that this therapy can obviously inhibit the growth of gastric cancer (Zhu et al. 2014b). Another study reported a novel application of adipose-derived mesenchymal stem cells (AD-MSCs)-based gene therapy activated using magnetic core-shell nanoparticles, and revealed a significant in vitro and in vivo ovarian cancer cell death (Yin et al. 2016).

3.7 Conclusion

During last decades the crucial question was gene therapy and stem cell therapy, which is more beneficial or superior to treatment? Gene therapy and stem cell therapy can complement each other. The goal of gene therapy is to correct a defective gene that causes the disease. In fact, the cause of the disease can be corrected with the help of gene therapy, but in stem cell therapy, the cell that is damaged is replaced by a healthy cell. In many cases, gene therapy and stem cell therapy are complementary because sometimes it is not possible to directly modify the gene in the living organism or in the cell, but it is necessary to construct cells that carry a healthy gene. In the last 7–8 years, researchers have succeeded in constructing cells that are called inducible stem cells (iPSCs). These cells are very similar to embryonic stem cells and have almost the same potential. The unique characteristics of inducible stem cells can be considered in this regard, which can be obtained from the patient himself. Therefore, the use of induction stem cells is a turning point in which therapeutic cell therapy and gene therapy come together so that they can be used to treat the disease. Up to now, patient-derived iPSCs that means carry their own genetic mutations have been confirmed to be invaluable to model genetic diseases at a cellular level, but still remains controversial whether similarly can perform to more complex pathophysiology diseases such as cancer as well. Researchers primarily interest cellular and molecular mechanisms, and in the end, to identify the agents and genes involved in the diseases. In the next step, and sometimes also in parallel, they attempted to replace the defective genes using stem cells along with gene therapy. Ultimately, all efforts in these researches are needed to better understand biological processes and help cure diseases.

References

Abrate A, Buono R, Canu T, Esposito A, Del Maschio A et al (2014) Mesenchymal stem cells expressing therapeutic genes induce autochthonous prostate tumour regression. Eur J Cancer 50:2478–2488

Ahmed AU, Thaci B, Alexiades NG, Han Y, Qian S et al (2011) Neural stem cell-based cell carriers enhance therapeutic efficacy of an oncolytic adenovirus in an orthotopic mouse model of human glioblastoma. Mol Ther 19:1714–1726

Ahn J, Lee H, Seo K, Kang S, Ra J, Youn H (2013) Anti-tumor effect of adipose tissue derived-mesenchymal stem cells expressing interferon-beta and treatment with cisplatin in a xenograft mouse model for canine melanoma. PLoS One 8:e74897

Alexiades NG, Auffinger B, Kim CK, Hasan T, Lee G et al (2015) MMP14 as a novel downstream target of VEGFR2 in migratory glioma-tropic neural stem cells. Stem Cell Res 15:598–607

Amano S, Li S, Gu C, Gao Y, Koizumi S et al (2009) Use of genetically engineered bone marrow-derived mesenchymal stem cells for glioma gene therapy. Int J Oncol 35:1265–1270

Amano S, Gu C, Koizumi S, Tokuyama T, Namba H (2011) Timing of ganciclovir administration in glioma gene therapy using HSVtk gene-transduced mesenchymal stem cells. Cancer Genomics Proteomics 8:245–250

Amato MP, Derfuss T, Hemmer B, Liblau R, Montalban X et al (2017) Environmental modifiable risk factors for multiple sclerosis: report from the 2016 ECTRIMS focused workshop Mult Scler 1352458516686847

Ardeshiry Lajimi A, Hagh MF, Saki N, Mortaz E, Soleimani M, Rahim F (2013) Feasibility of cell therapy in multiple sclerosis: a systematic review of 83 studies. Int J Hematol Oncol Stem Cell Res 7:15–33

Avior Y, Sagi I, Benvenisty N (2016) Pluripotent stem cells in disease modelling and drug discovery. Nat Rev Mol Cell Biol 17:170–182

Bahrambeigi V, Ahmadi N, Salehi R, Javanmard SH (2015) Genetically modified murine adipose-derived mesenchymal stem cells producing interleukin-2 favor B16F10 melanoma cell proliferation. Immunol Investig 44:216–236

Bailer SM, Funk C, Riedl A, Ruzsics Z (2017) Herpesviral vectors and their application in oncolytic therapy, vaccination, and gene transfer. Virus Genes 53(5):741–748

Ballios BG, van der Kooy D (2010) Biology and therapeutic potential of adult retinal stem cells. Can J Ophthalmol 45:342–351

Barkho BZ, Zhao X (2011) Adult neural stem cells: response to stroke injury and potential for therapeutic applications. Curr Stem Cell Res Ther 6:327–338

Bauer TR Jr, Allen JM, Hai M, Tuschong LM, Khan IF et al (2008) Successful treatment of canine leukocyte adhesion deficiency by foamy virus vectors. Nat Med 14:93–97

Bera A, Sen D (2017) Promise of adeno-associated virus as a gene therapy vector for cardiovascular diseases. Heart Fail Rev 22:795–823

Bexell D, Gunnarsson S, Tormin A, Darabi A, Gisselsson D et al (2009) Bone marrow multipotent mesenchymal stroma cells act as pericyte-like migratory vehicles in experimental gliomas. Mol Ther 17:183–190

Bexell D, Gunnarsson S, Svensson A, Tormin A, Henriques-Oliveira C et al (2012) Rat multipotent mesenchymal stromal cells lack long-distance tropism to 3 different rat glioma models. Neurosurgery 70:731–739

Bigger BW, Siapati EK, Mistry A, Waddington SN, Nivsarkar MS et al (2006) Permanent partial phenotypic correction and tolerance in a mouse model of hemophilia B by stem cell gene delivery of human factor IX. Gene Ther 13:117–126
Blackadar CB (2016) Historical review of the causes of cancer. World J Clin Oncol 7:54–86
Broxmeyer HE (2011) Insights into the biology of cord blood stem/progenitor cells. Cell Prolif 44(Suppl 1):55–59
Chang DY, Yoo SW, Hong Y, Kim S, Kim SJ et al (2010) The growth of brain tumors can be suppressed by multiple transplantation of mesenchymal stem cells expressing cytosine deaminase. Int J Cancer 127:1975–1983
Choi SA, Lee JY, Wang KC, Phi JH, Song SH et al (2012) Human adipose tissue-derived mesenchymal stem cells: characteristics and therapeutic potential as cellular vehicles for prodrug gene therapy against brainstem gliomas. Eur J Cancer 48:129–137
Choi SA, Lee YE, Kwak PA, Lee JY, Kim SS et al (2015) Clinically applicable human adipose tissue-derived mesenchymal stem cells delivering therapeutic genes to brainstem gliomas. Cancer Gene Ther 22:302–311
Christopher PR, Kingsley PA, Singh Bedi H, Singh Kwatra K, Rathore S, Das KC (2015) Large mid-esophageal granular cell tumor: benign versus malignant. Rare Tumors 7:5772
Connelly JP, Kwon EM, Gao Y, Trivedi NS, Elkahloun AG et al (2014) Targeted correction of RUNX1 mutation in FPD patient-specific induced pluripotent stem cells rescues megakaryopoietic defects. Blood 124:1926–1930
Cuenca N, Fernandez-Sanchez L, Campello L, Maneu V, De la Villa P et al (2014) Cellular responses following retinal injuries and therapeutic approaches for neurodegenerative diseases. Prog Retin Eye Res 43:17–75
Dembinski JL, Wilson SM, Spaeth EL, Studeny M, Zompetta C et al (2013) Tumor stroma engraftment of gene-modified mesenchymal stem cells as anti-tumor therapy against ovarian cancer. Cytotherapy 15:20–32
Deng YJ, Zhang SL, Shang XY, Fu PF, Cui XY (2005) [Transfection and expression of hRI gene on human umbilical blood stem cells and gene therapy for mouse melanoma]. Sheng wu gong cheng xue bao 21:36–41
Deng Q, Zhang Z, Feng X, Li T, Liu N et al (2014) TRAIL-secreting mesenchymal stem cells promote apoptosis in heat-shock-treated liver cancer cells and inhibit tumor growth in nude mice. Gene Ther 21:317–327
Deyle DR, Khan IF, Ren G, Wang PR, Kho J et al (2012) Normal collagen and bone production by gene-targeted human osteogenesis imperfecta iPSCs. Mol Ther 20:204–213
Dutta UR, Rajitha P, Pidugu VK, Dalal AB (2011) Cytogenetic abnormalities in 1162 couples with recurrent miscarriages in southern region of India: report and review. J Assist Reprod Genet 28:145–149
Dwyer RM, Ryan J, Havelin RJ, Morris JC, Miller BW et al (2011) Mesenchymal stem cell-mediated delivery of the sodium iodide symporter supports radionuclide imaging and treatment of breast cancer. Stem Cells 29:1149–1157
Earls JK, Jin S, Ye K (2013) Mechanobiology of human pluripotent stem cells. Tissue Eng B Rev 19:420–430
Ebrahimi A, Rahim F (2014) Recent immunomodulatory strategies in transplantation. Immunol Investig 43:829–837
Ebrahimi A, Hosseini SA, Rahim F (2014) Immunosuppressive therapy in allograft transplantation: from novel insights and strategies to tolerance and challenges. Cent Eur J Immunol 39:400–409
Elmore S (2007) Apoptosis: a review of programmed cell death. Toxicol Pathol 35:495–516
Emery DW (2004) Gene therapy for genetic diseases: on the horizon. Clin Appl Immunol Rev 4:411–422
Farajkhoda T (2017) An overview on ethical considerations in stem cell research in Iran and ethical recommendations: a review. Int J Reprod Biomed 15:67–74
Fareed M, Afzal M (2017) Genetics of consanguinity and inbreeding in health and disease. Ann Hum Biol 44:99–107

Fei S, Qi X, Kedong S, Guangchun J, Jian L, Wei Q (2012) The antitumor effect of mesenchymal stem cells transduced with a lentiviral vector expressing cytosine deaminase in a rat glioma model. J Cancer Res Clin Oncol 138:347–357

Feng R, Hu J, Wu J, Lang L, Ma C et al (2017) Comprehensive preoperative work-up and surgical treatment of low grade tumor/benign lesion related temporal lobe epilepsy. J Clin Neurosci 39:203–208

Firth AL, Menon T, Parker GS, Qualls SJ, Lewis BM et al (2015) Functional gene correction for cystic fibrosis in lung epithelial cells generated from patient iPSCs. Cell Rep 12:1385–1390

Flynn R, Grundmann A, Renz P, Hanseler W, James WS et al (2015) CRISPR-mediated genotypic and phenotypic correction of a chronic granulomatous disease mutation in human iPS cells. Exp Hematol 43:838–48.e3

Fougeroux C, Holst PJ (2017) Future prospects for the development of cost-effective adenovirus vaccines. Int J Mol Sci 18(4):686

Galeev R, Karlsson C, Baudet A, Larsson J (2017) Forward RNAi screens in human hematopoietic stem cells. Methods Mol Biol 1622:29–50

Gamez A, Yuste-Checa P, Brasil S, Briso-Montiano A, Desviat LR et al (2018) Protein misfolding diseases: prospects of pharmacological treatment. Clin Genet 93(3):450–458

Gammaitoni L, Lucchi S, Bruno S, Tesio M, Gunetti M et al (2006) Serial transplantations in nonobese diabetic/severe combined immunodeficiency mice of transduced human CD34+ cord blood cells: efficient oncoretroviral gene transfer and ex vivo expansion under serum-free conditions. Stem Cells 24:1201–1212

Gao P, Ding Q, Wu Z, Jiang H, Fang Z (2010a) Therapeutic potential of human mesenchymal stem cells producing IL-12 in a mouse xenograft model of renal cell carcinoma. Cancer Lett 290:157–166

Gao Y, Yao A, Zhang W, Lu S, Yu Y et al (2010b) Human mesenchymal stem cells overexpressing pigment epithelium-derived factor inhibit hepatocellular carcinoma in nude mice. Oncogene 29:2784–2794

Gao Y, Guo X, Santostefano K, Wang Y, Reid T et al (2016) Genome therapy of myotonic dystrophy type 1 iPS cells for development of autologous stem cell therapy. Mol Ther 24:1378–1387

Hakkarainen T, Sarkioja M, Lehenkari P, Miettinen S, Ylikomi T et al (2007) Human mesenchymal stem cells lack tumor tropism but enhance the antitumor activity of oncolytic adenoviruses in orthotopic lung and breast tumors. Hum Gene Ther 18:627–641

Haley K (2017) Congenital hemolytic anemia. Med Clin North Am 101:361–374

Hoyos V, Del Bufalo F, Yagyu S, Ando M, Dotti G et al (2015) Mesenchymal stromal cells for linked delivery of oncolytic and apoptotic adenoviruses to non-small-cell lung cancers. Mol Ther 23:1497–1506

Hu W, Wang J, He X, Zhang H, Yu F et al (2011) Human umbilical blood mononuclear cell-derived mesenchymal stem cells serve as interleukin-21 gene delivery vehicles for epithelial ovarian cancer therapy in nude mice. Biotechnol Appl Biochem 58:397–404

Iizuka H, Kagoya Y, Kataoka K, Yoshimi A, Miyauchi M et al (2015) Targeted gene correction of RUNX1 in induced pluripotent stem cells derived from familial platelet disorder with propensity to myeloid malignancy restores normal megakaryopoiesis. Exp Hematol 43:849–857

Ikezawa M, Cao B, Qu Z, Peng H, Xiao X et al (2003) Dystrophin delivery in dystrophin-deficient DMDmdx skeletal muscle by isogenic muscle-derived stem cell transplantation. Hum Gene Ther 14:1535–1546

Jayant RD, Sosa D, Kaushik A, Atluri V, Vashist A et al (2016) Current status of non-viral gene therapy for CNS disorders. Expert Opin Drug Deliv 13:1433–1445

Jeong KY, Lee EJ, Kim SJ, Yang SH, Sung YC, Seong J (2015) Irradiation-induced localization of IL-12-expressing mesenchymal stem cells to enhance the curative effect in murine metastatic hepatoma. Int J Cancer 137:721–730

Jiang J, Chen W, Zhuang R, Song T, Li P (2010) The effect of endostatin mediated by human mesenchymal stem cells on ovarian cancer cells in vitro. J Cancer Res Clin Oncol 136:873–881

Jin G, Zhou Y, Chai Q, Zhu G, Xu F, Liu F (2013) VP22 and cytosine deaminase fusion gene modified tissue-engineered neural stem cells for glioma therapy. J Cancer Res Clin Oncol 139:475–483

Joo KM, Park IH, Shin JY, Jin J, Kang BG et al (2009) Human neural stem cells can target and deliver therapeutic genes to breast cancer brain metastases. Mol Ther 17:570–575

Kanehira M, Xin H, Hoshino K, Maemondo M, Mizuguchi H et al (2007) Targeted delivery of NK4 to multiple lung tumors by bone marrow-derived mesenchymal stem cells. Cancer Gene Ther 14:894–903

Kaneti L, Bronshtein T, Malkah Dayan N, Kovregina I, Letko Khait N et al (2016) Nanoghosts as a novel natural nonviral gene delivery platform safely targeting multiple cancers. Nano Lett 16:1574–1582

Kaufmann KB, Büning H, Galy A, Schambach A, Grez M (2013) Gene therapy on the move. EMBO Mol Med 5:1642–1661

Khanna S, Gharpure AS (2017) Petroleum carcinogenicity and aerodigestive tract: in context of developing nations. Cureus 9:e1202

Kim SK, Kim SU, Park IH, Bang JH, Aboody KS et al (2006) Human neural stem cells target experimental intracranial medulloblastoma and deliver a therapeutic gene leading to tumor regression. Clin Cancer Res 12:5550–5556

Kim KY, Kim SU, Leung PC, Jeung EB, Choi KC (2010) Influence of the prodrugs 5-fluorocytosine and CPT-11 on ovarian cancer cells using genetically engineered stem cells: tumor-tropic potential and inhibition of ovarian cancer cell growth. Cancer Sci 101:955–962

Kim JH, Kim JY, Kim SU, Cho KG (2012a) Therapeutic effect of genetically modified human neural stem cells encoding cytosine deaminase on experimental glioma. Biochem Biophys Res Commun 417:534–540

Kim KY, Yi BR, Lee HR, Kang NH, Jeung EB et al (2012b) Stem cells with fused gene expression of cytosine deaminase and interferon-beta migrate to human gastric cancer cells and result in synergistic growth inhibition for potential therapeutic use. Int J Oncol 40:1097–1104

Kim SM, Woo JS, Jeong CH, Ryu CH, Lim JY, Jeun SS (2012c) Effective combination therapy for malignant glioma with TRAIL-secreting mesenchymal stem cells and lipoxygenase inhibitor MK886. Cancer Res 72:4807–4817

Kim HS, Yi BR, Hwang KA, Kim SU, Choi KC (2013a) Anticancer effects of the engineered stem cells transduced with therapeutic genes via a selective tumor tropism caused by vascular endothelial growth factor toward HeLa cervical cancer cells. Mol Cells 36:347–354

Kim SW, Kim SJ, Park SH, Yang HG, Kang MC et al (2013b) Complete regression of metastatic renal cell carcinoma by multiple injections of engineered mesenchymal stem cells expressing dodecameric TRAIL and HSV-TK. Clin Cancer Res 19:415–427

Kim JH, Lee HJ, Song YS (2014a) Stem cell based gene therapy in prostate cancer. Biomed Res Int 2014:549136

Kim SM, Woo JS, Jeong CH, Ryu CH, Jang JD, Jeun SS (2014b) Potential application of temozolomide in mesenchymal stem cell-based TRAIL gene therapy against malignant glioma. Stem Cells Transl Med 3:172–182

Knoop K, Schwenk N, Schmohl K, Muller A, Zach C et al (2015) Mesenchymal stem cell-mediated, tumor stroma-targeted radioiodine therapy of metastatic colon cancer using the sodium iodide symporter as theranostic gene. J Nucl Med 56:600–606

Kosaka H, Ichikawa T, Kurozumi K, Kambara H, Inoue S et al (2012) Therapeutic effect of suicide gene-transferred mesenchymal stem cells in a rat model of glioma. Cancer Gene Ther 19:572–578

Krasikova LS, Karshieva SS, Cheglakov IB, Belyavsky AV (2015) [Mesenchymal stem cells expressing cytosine deaminase inhibit growth of murine melanoma B16F10 in vivo]. Mol Biol 49:1007–1015

Kucerova L, Skolekova S, Demkova L, Bohovic R, Matuskova M (2014) Long-term efficiency of mesenchymal stromal cell-mediated CD-MSC/5FC therapy in human melanoma xenograft model. Gene Ther 21:874–887

Kugler J, Huhse B, Tralau T, Luch A (2017) Embryonic stem cells and the next generation of developmental toxicity testing. Expert Opin Drug Metab Toxicol 13:833–841

Lakota J, Gocarova K, Spanik S (2015) Treatment of metastatic head and neck cancer with mesenchymal stem cells combined with prodrug gene therapy. Exp Oncol 37:298

Lau AA, Hemsley KM (2017) Adeno-associated viral gene therapy for mucopolysaccharidoses exhibiting neurodegeneration. J Mol Med 95(10):1043–1052
Lee KY (2017) A unified pathogenesis for kidney diseases, including genetic diseases and cancers, by the protein-homeostasis-system hypothesis. Kidney Res Clin Pract 36:132–144
Lee HJ, Yang HM, Choi YS, Park SH, Moon SH et al (2013) A therapeutic strategy for metastatic malignant fibrous histiocytoma through mesenchymal stromal cell-mediated TRAIL production. Ann Surg 257:952–960
Leten C, Trekker J, Struys T, Dresselaers T, Gijsbers R et al (2015) Assessment of bystander killing-mediated therapy of malignant brain tumors using a multimodal imaging approach. Stem Cell Res Ther 6:163
Li X, Lu Y, Huang W, Xu H, Chen X et al (2006) In vitro effect of adenovirus-mediated human Gamma Interferon gene transfer into human mesenchymal stem cells for chronic myelogenous leukemia. Hematol Oncol 24:151–158
Li L, Li S, Cai T, Wang H, Xie X et al (2016) The targeted inhibitory effects of human amniotic fluid stem cells carrying CXCR4 promoter and DAL-1 on non-small cell lung carcinoma growth. Gene Ther 23:214–222
Liu YN, Yan X, Sun Z, Han Q, Zhao RC (2007) [Mice adipose derived Flk-1+ mesenchymal stem cells can ameliorate Duchenne's muscular dystrophy in Mdx mice for their multilineage potential]. Zhongguo Shi Yan Xue Ye Xue Za Zhi 15:306–312
Long Q, Yang R, Lu W, Zhu W, Zhou J et al (2017) Adenovirus-mediated truncated Bid overexpression induced by the Cre/LoxP system promotes the cell apoptosis of CD133+ ovarian cancer stem cells. Oncol Rep 37:155–162
Lopez-Ornelas A, Vergara P, Segovia J (2014) Neural stem cells producing an inducible and soluble form of Gas1 target and inhibit intracranial glioma growth. Cytotherapy 16:1011–1023
Ma GL, Zhu XH, Xuan SY, Mao WZ (2012) [Study of inhibiting and killing effects of transgenic LIGHT human umbilical cord blood mesenchymal stem cells on stomach cancer]. Zhonghua wei chang wai ke za zhi 15:1178–1181
Mac Gabhann F, Annex BH, Popel AS (2010) Gene therapy from the perspective of systems biology. Curr Opin Mol Ther 12:570–577
Mao W, Zhu X, Tang D, Zhao Y, Zhao B et al (2012) TNF-alpha expression in the UCB-MSCs as stable source inhibits gastric cancers growth in nude mice. Cancer Investig 30:463–472
Marsico C, Kimberlin DW (2017) Congenital cytomegalovirus infection: advances and challenges in diagnosis, prevention and treatment. Ital J Pediatr 43:38
Masiuk KE, Brown D, Laborada J, Hollis RP, Urbinati F, Kohn DB (2017) Improving gene therapy efficiency through the enrichment of human hematopoietic stem cells. Mol Ther 25(9):2163–2175
Mason EF, Rathmell JC (2011) Cell metabolism: an essential link between cell growth and apoptosis. Biochim Biophys Acta 1813:645–654
Matuskova M, Kozovska Z, Toro L, Durinikova E, Tyciakova S et al (2015) Combined enzyme/prodrug treatment by genetically engineered AT-MSC exerts synergy and inhibits growth of MDA-MB-231 induced lung metastases. J Exp Clin Cancer Res 34:33
Mavroudi M, Zarogoulidis P, Porpodis K, Kioumis I, Lampaki S et al (2014) Stem cells' guided gene therapy of cancer: new frontier in personalized and targeted therapy. J Cancer Res Ther 2:22–33
Menon T, Firth AL, Scripture-Adams DD, Galic Z, Qualls SJ et al (2015) Lymphoid regeneration from gene-corrected SCID-X1 subject-derived iPSCs. Cell Stem Cell 16:367–372
Mohammadian F, Eatemadi A, Daraee H (2017) Inner ear drug delivery using liposomes. Cell Mol Biol 63:28–33
Momtaz R, Ghanem NM, El-Makky NM, Ismail MA (2017) Integrated analysis of SNP, CNV and gene expression data in genetic association studies. Clin Genet 93(3):557–566
Moniri MR, Sun XY, Rayat J, Dai D, Ao Z et al (2012) TRAIL-engineered pancreas-derived mesenchymal stem cells: characterization and cytotoxic effects on pancreatic cancer cells. Cancer Gene Ther 19:652–658

Morgan JE, Partridge TA (1992) Cell transplantation and gene therapy in muscular dystrophy. Bioessays 14:641–645
Murai J (2017) Targeting DNA repair and replication stress in the treatment of ovarian cancer. Int J Clin Oncol 22(4):619–628
Nakamura K, Ito Y, Kawano Y, Kurozumi K, Kobune M et al (2004) Antitumor effect of genetically engineered mesenchymal stem cells in a rat glioma model. Gene Ther 11:1155–1164
Naso MF, Tomkowicz B, Perry WL III, Strohl WR (2017) Adeno-associated virus (AAV) as a vector for gene therapy. BioDrugs 31(4):317–334
Navarro S, Moleiro V, Molina-Estevez FJ, Lozano ML, Chinchon R et al (2014) Generation of iPSCs from genetically corrected Brca2 hypomorphic cells: implications in cell reprogramming and stem cell therapy. Stem Cells 32:436–446
Naylor S, Chen JY (2010) Unraveling human complexity and disease with systems biology and personalized medicine. Pers Med 7:275–289
Niess H, Bao Q, Conrad C, Zischek C, Notohamiprodjo M et al (2011) Selective targeting of genetically engineered mesenchymal stem cells to tumor stroma microenvironments using tissue-specific suicide gene expression suppresses growth of hepatocellular carcinoma. Ann Surg 254:767–774; discussion 74–5
Niu X, He W, Song B, Ou Z, Fan D et al (2016) Combining single strand oligodeoxynucleotides and CRISPR/Cas9 to correct gene mutations in beta-thalassemia-induced pluripotent stem cells. J Biol Chem 291:16576–16585
O'Neill AC, Jagannathan JP, Ramaiya NH (2017) Evolving cancer classification in the era of personalized medicine: a primer for radiologists. Korean J Radiol 18:6–17
Ochakovski GA, Bartz-Schmidt KU, Fischer MD (2017) Retinal gene therapy: surgical vector delivery in the translation to clinical trials. Front Neurosci 11:174
Oldfors A, Nordborg C, Rockert L, Parkkila T, Kalimo H (1988) [Primary benign lymphoma—tumor biology, pathology and clinical aspects]. Lakartidningen 85:1797, 99–800
Otsu M (2015) Perspectives on stem cell gene therapy for genetic disorders. ISBT Sci Ser 10:231–234
Otsuka A, Nomura T, Rerknimitr P, Seidel JA, Honda T, Kabashima K (2017) The interplay between genetic and environmental factors in the pathogenesis of atopic dermatitis. Immunol Rev 278:246–262
Pawliuk R, Bachelot T, Wise RJ, Mathews-Roth MM, Leboulch P (1999) Long-term cure of the photosensitivity of murine erythropoietic protoporphyria by preselective gene therapy. Nat Med 5:768–773
Porada CD, Almeida-Porada G (2010) Mesenchymal stem cells as therapeutics and vehicles for gene and drug delivery. Adv Drug Deliv Rev 62:1156–1166
Purwanti YI, Chen C, Lam DH, Wu C, Zeng J et al (2014) Antitumor effects of CD40 ligand-expressing endothelial progenitor cells derived from human induced pluripotent stem cells in a metastatic breast cancer model. Stem Cells Transl Med 3:923–935
Qazi Y, Hamrah P (2013) Gene therapy in corneal transplantation. Semin Ophthalmol 28:287–300
Rachakatla RS, Marini F, Weiss ML, Tamura M, Troyer D (2007) Development of human umbilical cord matrix stem cell-based gene therapy for experimental lung tumors. Cancer Gene Ther 14:828–835
Rahim F, Allahmoradi H, Salari F, Shahjahani M, Fard AD et al (2013) Evaluation of signaling pathways involved in gamma-globin gene induction using fetal hemoglobin inducer drugs. Int J Hematol Oncol Stem Cell Res 7:41–46
Ramnaraine ML, Mathews WE, Clohisy DR (2012) Lentivirus transduction of human osteoclast precursor cells and differentiation into functional osteoclasts. Bone 50:97–103
Rappaport SM (2016) Genetic factors are not the major causes of chronic diseases. PLoS One 11:e0154387
Ren C, Kumar S, Chanda D, Kallman L, Chen J et al (2008) Cancer gene therapy using mesenchymal stem cells expressing interferon-beta in a mouse prostate cancer lung metastasis model. Gene Ther 15:1446–1453

Riley MK, Vermerris W (2017) Recent advances in nanomaterials for gene delivery—a review. Nanomaterials 7
Roquette R, Painho M, Nunes B (2017) Spatial epidemiology of cancer: a review of data sources, methods and risk factors. Geospat Health 12:504
Rouanet M, Lebrin M, Gross F, Bournet B, Cordelier P, Buscail L (2017) Gene therapy for pancreatic cancer: specificity, issues and hopes. Int J Mol Sci 18:1231
Rubanyi GM (2001) The future of human gene therapy. Mol Asp Med 22:113–142
Ryu CH, Park KY, Kim SM, Jeong CH, Woo JS et al (2012) Valproic acid enhances anti-tumor effect of mesenchymal stem cell mediated HSV-TK gene therapy in intracranial glioma. Biochem Biophys Res Commun 421:585–590
Saffari M, Moghimi HR, Dass CR (2016) Barriers to liposomal gene delivery: from application site to the target. Iran J Pharm Res 15:3–17
Saki N, Jalalifar MA, Soleimani M, Hajizamani S, Rahim F (2013) Adverse effect of high glucose concentration on stem cell therapy. Int J Hematol Oncol Stem Cell Res 7:34–40
Sampaolesi M, Blot S, D'Antona G, Granger N, Tonlorenzi R et al (2006) Mesoangioblast stem cells ameliorate muscle function in dystrophic dogs. Nature 444:574–579
Schakman O, Thissen JP (2006) Gene therapy with anabolic growth factors to prevent muscle atrophy. Curr Opin Clin Nutr Metab Care 9:207–213
Seo KW, Lee HW, Oh YI, Ahn JO, Koh YR et al (2011) Anti-tumor effects of canine adipose tissue-derived mesenchymal stromal cell-based interferon-beta gene therapy and cisplatin in a mouse melanoma model. Cytotherapy 13:944–955
Seol HJ, Jin J, Seong DH, Joo KM, Kang W et al (2011) Genetically engineered human neural stem cells with rabbit carboxyl esterase can target brain metastasis from breast cancer. Cancer Lett 311:152–159
Shahrabi S, Azizidoost S, Shahjahani M, Rahim F, Ahmadzadeh A, Saki N (2014) New insights in cellular and molecular aspects of BM niche in chronic myelogenous leukemia. Tumour Biol 35:10627–10633
Shawky RM, Elsayed NS, Ibrahim DS, Seifeldin NS (2012) Profile of genetic disorders prevalent in northeast region of Cairo, Egypt. Egypt J Med Hum Genet 13:45–62
Sheppard M, Spencer RN, Ashcroft R, David AL (2016) Ethics and social acceptability of a proposed clinical trial using maternal gene therapy to treat severe early-onset fetal growth restriction. Ultrasound Obstet Gynecol 47:484–491
Siapati EK, Bigger BW, Kashofer K, Themis M, Thrasher AJ, Bonnet D (2007) Murine leukemia following irradiation conditioning for transplantation of lentivirally-modified hematopoietic stem cells. Eur J Haematol 78:303–313
Sideri A, Neokleous N, Brunet De La Grange P, Guerton B, Le Bousse Kerdilles MC et al (2011) An overview of the progress on double umbilical cord blood transplantation. Haematologica 96:1213–1220
Skokowa J, Dale DC, Touw IP, Zeidler C, Welte K (2017) Severe congenital neutropenias. Nat Rev Dis Primers 3:17032
Song P, He J, Li F, Jin C (2017) Innovative measures to combat rare diseases in China: the national rare diseases registry system, larger-scale clinical cohort studies, and studies in combination with precision medicine research. Intractable Rare Dis Res 6:1–5
Stoff-Khalili MA, Rivera AA, Mathis JM, Banerjee NS, Moon AS et al (2007) Mesenchymal stem cells as a vehicle for targeted delivery of CRAds to lung metastases of breast carcinoma. Breast Cancer Res Treat 105:157–167
Sudhakar A (2009) History of cancer, ancient and modern treatment methods. J Cancer Sci Ther 1:1–4
Sun N, Zhao H (2014) Seamless correction of the sickle cell disease mutation of the HBB gene in human induced pluripotent stem cells using TALENs. Biotechnol Bioeng 111:1048–1053
Tedesco FS, Gerli MF, Perani L, Benedetti S, Ungaro F et al (2012) Transplantation of genetically corrected human iPSC-derived progenitors in mice with limb-girdle muscular dystrophy. Sci Transl Med 4:140ra89

Teng J, Hejazi S, Badr CE, Tannous BA (2014) Systemic anticancer neural stem cells in combination with a cardiac glycoside for glioblastoma therapy. Stem Cells 32:2021–2032

Traebert J, Schneider IJ, Colussi CF, de Lacerda JT (2013) Burden of disease due to cancer in a Southern Brazilian state. Cancer Epidemiol 37:788–792

Tweedell KS (2017) The adaptability of somatic stem cells: a review. J Stem Cells Regen Med 13:3–13

Tyciakova S, Matuskova M, Bohovic R, Polakova K, Toro L et al (2015) Genetically engineered mesenchymal stromal cells producing TNFalpha have tumour suppressing effect on human melanoma xenograft. J Gene Med 17:54–67

Uzzaman M, Benveniste RJ, Keller G, Germano IM (2005) Embryonic stem cell-derived astrocytes: a novel gene therapy vector for brain tumors. Neurosurg Focus 19:E6

Wang Y, Zheng CG, Jiang Y, Zhang J, Chen J et al (2012) Genetic correction of beta-thalassemia patient-specific iPS cells and its use in improving hemoglobin production in irradiated SCID mice. Cell Res 22:637–648

Wang G, Yuan N, Huang S, Feng L, Han R et al (2017) The CNGRCLLII(KLAKLAK)2 peptide shows cytotoxicity against HUVECs by inducing apoptosis: an in vitro and in vivo study. Tumour Biol 39:1010428317701649

Warnock JN, Daigre C, Al-Rubeai M (2011) Introduction to viral vectors. Methods Mol Biol 737:1–25

Warrick E, Garcia M, Chagnoleau C, Chevallier O, Bergoglio V et al (2012) Preclinical corrective gene transfer in xeroderma pigmentosum human skin stem cells. Mol Ther 20:798–807

Williams DA (2014) Curing genetic disease with gene therapy. Trans Am Clin Climatol Assoc 125:122–129

Williams DA, Croop J, Kelly P (2005) Gene therapy in the treatment of Fanconi anemia, a progressive bone marrow failure syndrome. Curr Opin Mol Ther 7:461–466

Wirth T (2011) A short perspective on gene therapy: clinical experience on gene therapy of gliomablastoma multiforme. World J Exp Med 1:10–16

Witt R, MacKenzie TC, Peranteau WH (2017) Fetal stem cell and gene therapy. Semin Fetal Neonatal Med 22(6):410–414

Wu Y, Zhou H, Fan X, Zhang Y, Zhang M et al (2015) Correction of a genetic disease by CRISPR-Cas9-mediated gene editing in mouse spermatogonial stem cells. Cell Res 25:67–79

Xin H, Kanehira M, Mizuguchi H, Hayakawa T, Kikuchi T et al (2007) Targeted delivery of CX3CL1 to multiple lung tumors by mesenchymal stem cells. Stem Cells 25:1618–1626

Xu G, Guo Y, Seng Z, Cui G, Qu J (2015) Bone marrow-derived mesenchymal stem cells co-expressing interleukin-18 and interferon-beta exhibit potent antitumor effect against intracranial glioma in rats. Oncol Rep 34:1915–1922

Yan C, Li S, Li Z, Peng H, Yuan X et al (2013) Human umbilical cord mesenchymal stem cells as vehicles of CD20-specific TRAIL fusion protein delivery: a double-target therapy against non-Hodgkin's lymphoma. Mol Pharm 10:142–151

Yang B, Qiu Y, Zhou N, Ouyang H, Ding J et al (2017) Application of stem cells in oral disease therapy: progresses and perspectives. Front Physiol 8:197

Yap NY, Rajandram R, Ng KL, Pailoor J, Fadzli A, Gobe GC (2015) Genetic and chromosomal aberrations and their clinical significance in renal neoplasms. Biomed Res Int 2015:476508

Yi BR, O SN, Kang NH, Hwang KA, Kim SU et al (2011) Genetically engineered stem cells expressing cytosine deaminase and interferon-beta migrate to human lung cancer cells and have potentially therapeutic anti-tumor effects. Int J Oncol 39:833–839

Yi BR, Park MA, Lee HR, Kang NH, Choi KJ et al (2013) Suppression of the growth of human colorectal cancer cells by therapeutic stem cells expressing cytosine deaminase and interferon-beta via their tumor-tropic effect in cellular and xenograft mouse models. Mol Oncol 7:543–554

Yi BR, Kim SU, Choi KC (2015) Additional effects of engineered stem cells expressing a therapeutic gene and interferon-beta in a xenograft mouse model of endometrial cancer. Int J Oncol 47:171–178

Yin PT, Shah S, Pasquale NJ, Garbuzenko OB, Minko T, Lee KB (2016) Stem cell-based gene therapy activated using magnetic hyperthermia to enhance the treatment of cancer. Biomaterials 81:46–57

Yoshida T, Ohnami S, Aoki K (2004) Development of gene therapy to target pancreatic cancer. Cancer Sci 95:283–289

Yoshimura N, Kato R, Chancellor MB, Nelson JB, Glorioso JC (2010) Gene therapy as future treatment of erectile dysfunction. Expert Opin Biol Ther 10:1305–1314

You Q, Yao Y, Zhang Y, Fu S, Du M, Zhang G (2015) Effect of targeted ovarian cancer therapy using amniotic fluid mesenchymal stem cells transfected with enhanced green fluorescent protein-human interleukin-2 in vivo. Mol Med Rep 12:4859–4866

Zhang QS (2017) Stem cell therapy for Fanconi anemia. Adv Exp Med Biol 1–10

Zhang S, Chen S, Li W, Guo X, Zhao P et al (2011) Rescue of ATP7B function in hepatocyte-like cells from Wilson's disease induced pluripotent stem cells using gene therapy or the chaperone drug curcumin. Hum Mol Genet 20:3176–3187

Zhang B, Shan H, Li D, Li ZR, Zhu KS, Jiang ZB (2012) The inhibitory effect of MSCs expressing TRAIL as a cellular delivery vehicle in combination with cisplatin on hepatocellular carcinoma. Cancer Biol Ther 13:1175–1184

Zhang D, Zheng L, Shi H, Chen X, Wan Y et al (2014a) Suppression of peritoneal tumorigenesis by placenta-derived mesenchymal stem cells expressing endostatin on colorectal cancer. Int J Med Sci 11:870–879

Zhang TY, Huang B, Yuan ZY, Hu YL, Tabata Y, Gao JQ (2014b) Gene recombinant bone marrow mesenchymal stem cells as a tumor-targeted suicide gene delivery vehicle in pulmonary metastasis therapy using non-viral transfection. Nanomedicine 10:257–267

Zhang Y, Wang J, Ren M, Li M, Chen D et al (2014c) Gene therapy of ovarian cancer using IL-21-secreting human umbilical cord mesenchymal stem cells in nude mice. J Ovarian Res 7:8

Zhang J, Hou L, Wu X, Zhao D, Wang Z et al (2016) Inhibitory effect of genetically engineered mesenchymal stem cells with Apoptin on hepatoma cells in vitro and in vivo. Mol Cell Biochem 416:193–203

Zhao WH, Cheng JX, Shi PF, Huang JY (2011) [Human umbilical cord mesenchymal stem cells with adenovirus-mediated interleukin 12 gene transduction inhibits the growth of ovarian carcinoma cells both in vitro and in vivo]. Nan fang yi ke da xue xue bao 31:903–907

Zhao Y, Lam DH, Yang J, Lin J, Tham CK et al (2012) Targeted suicide gene therapy for glioma using human embryonic stem cell-derived neural stem cells genetically modified by baculoviral vectors. Gene Ther 19:189–200

Zhong XG, He S, Yin W, Deng JY, Chen B (2005) [Tropism of adult liver stem cells toward hepatocellular carcinoma cells in vitro]. Zhonghua gan zang bing za zhi = Zhonghua ganzangbing zazhi 13:644–647

Zhong XG, He S, Yin W, Deng JY, Cheng B (2007) Selective tropism of liver stem cells to hepatocellular carcinoma in vivo. World J Gastroenterol 13:3886–3891

Zhou Y, Abel GA, Hamilton W, Pritchard-Jones K, Gross CP et al (2017) Diagnosis of cancer as an emergency: a critical review of current evidence. Nat Rev Clin Oncol 14:45–56

Zhu X, Su D, Xuan S, Ma G, Dai Z et al (2013) Gene therapy of gastric cancer using LIGHT-secreting human umbilical cord blood-derived mesenchymal stem cells. Gastric Cancer 16:155–166

Zhu D, Chen C, Purwanti YI, Du S, Lam DH et al (2014a) Induced pluripotent stem cell-derived neural stem cells transduced with baculovirus encoding CD40 ligand for immunogene therapy in mouse models of breast cancer. Hum Gene Ther 25:747–758

Zhu Y, Cheng M, Yang Z, Zeng CY, Chen J et al (2014b) Mesenchymal stem cell-based NK4 gene therapy in nude mice bearing gastric cancer xenografts. Drug Des Devel Ther 8:2449–2462

Zischek C, Niess H, Ischenko I, Conrad C, Huss R et al (2009) Targeting tumor stroma using engineered mesenchymal stem cells reduces the growth of pancreatic carcinoma. Ann Surg 250:747–753

Zolochevska O, Yu G, Gimble JM, Figueiredo ML (2012) Pigment epithelial-derived factor and melanoma differentiation associated gene-7 cytokine gene therapies delivered by adipose-derived stromal/mesenchymal stem cells are effective in reducing prostate cancer cell growth. Stem Cells Dev 21:1112–1123

Chapter 4
Cancer Stem Cells (CSCs) and Their Application in Cancer Immunotherapy

Deepak Rohila, Yang Yao, and Xuexiao Jin

Abbreviations

ALDH	Aldehyde dehydrogenases
AML	Acute myelocytic leukemia
CSC	Cancer stem cell
EMT	Epithelial-mesenchymal transition
LGR5	Leucine-rich repeat-containing G-protein coupled receptor 5
SMO	Smoothened

4.1 Introduction

Twenty percent of all deaths in developed countries are caused by cancer and is thus emerging as one of the major health problems in the world. Common treatments involve surgery, radiotherapy, and chemotherapy. Although reoccurrence of tumor or secondary manifestation clearly indicates that some tumor cells can survive post chemo- and radiotherapy by activating their self-renewal pathways (Ahmed et al. 2013). Hence, researchers are always seeking for alternative strategies for resistive tumor, and in this context the cancer stem cell targeted therapy is one of these new strategies.

Cancer stem cells were initially described in studies of chronic myeloid leukemia (CML) which belongs to a type of blood cancer (Lapidot et al. 1994). The cancer stem cell (CSC) hypothesis posits the existence of subpopulations of cancer cells with indefinite proliferative abilities that drive the formation and metastasis of tumors.

D. Rohila · Y. Yao (✉) · X. Jin
Institute of Immunology, Zhejiang University School of Medicine, Hangzhou, China
e-mail: 0916036@zju.edu.cn

P. V. Pham, A. El-Hashash (eds.), *Stem Cells for Cancer and Genetic Disease Treatment*, Stem Cells in Clinical Applications, https://doi.org/10.1007/978-3-319-98065-2_4

4.2 Molecular Characterization of Cancer Stem Cells and Their Associated Markers

Cancer stem cells are well known for their stemness properties of self-renewal and differentiation as well as their ability for tumor initiation and conferring chemo/radiotherapy resistance by activating the self-renewal pathways (Lupia and Cavallaro 2017).

Stem cells are well known for their ability of self-renewal and generation of mature cells of any specific tissue type via differentiation. Notably, cancer cells share significant similarities with stem cells. One, similar to normal stem cells, cancer cells also have the ability of self-renewal. Two, the possibility of generation of tumor cells from stem cells, and three tumor cells might also contain some cancer stem cells—rare cells with indefinite proliferative properties for formation and growth of tumors (Eaves 2008). The cancer stem cells hypothesis proposes the existence of neoplastic sub-population cell residing in a tumor that has capability to divide asymmetrically, generating daughter cells.

Thus far, only a few markers have been used to identify CSC cells with certain tumors, like lung (Barr et al. 2017), ovarian (Ahmed et al. 2013), breast (Al-Dhfyan et al. 2017), colon (Bhatlekar et al. 2017), and prostate (Ceder and Elgqvist 2017).

Many reports have suggested that CSCs of different kinds of cancer may express specific molecules, which could be treated as CSCs markers (Xiao et al. 2017). The variety in CSC-associated markers can directly correlate with the existence of a separate sub-population of these cells residing inside the tumor tissues.

Table 4.1 provides a list of potential markers of cancer stem cells from different tissue origin (Maccalli et al. 2014; Medema 2013).

Table 4.1 CSC markers for distinct tumor types

Markers	Tumor type	References
CD34+CD38−	Acute myelocytic leukemia (AML)	Picot et al. (2017)
ALDH1+	Colon cancer, cervical cancer, non-small cell lung cancer (NSCLC), mesenchymal glioma	Holah et al. (2017), Sullivan et al. (2017), Wang et al. (2017b, c)
CD133+	Laryngeal squamous cell carcinoma	Wang et al. (2017a)
LGR5+	Colon cancer, gastric cancer, skin squamous cell carcinoma	Barker et al. (2007), Stigbrand (2017), Yanai et al. (2017), Zhang et al. (2017) Melo et al. (2017)
CD44+ CD24-	Breast cancer	Singh et al. (2017)
CD24+	Colon cancer	Ke et al. (2012)
CD166+	Colon cancer; non-small cell lung cancer (NSCLC)	Margaritescu et al. (2014), Tachezy et al. (2014)
CD90+	Liver cancer, esophageal squamous cell carcinoma	Wang et al. (2017d), Yang et al. (2016)
CD114+	Melanoma, neuroblastoma	Zage et al. (2017)

The CSC populations vary among tumors and contribute to only 0.1–10% of all the tumor cells, and CSC markers that are highly expressed in CSCs may not be highly expressed in tumors. Although these facts make it challenging to identify unique molecular markers of CSCs, the endeavor of characterizing these markers still suggests a good start.

4.3 Signaling Pathways Involved in Therapy Resistance of CSCs

Thanks to the advances in modern research tools, it has been possible to reveal various signal transduction pathways that play a key role in cancer stemness. Alteration in the notch, wnt, and sonic hedgehog pathways gives insight indication about the factors that activated in cancer via signals that promote stem cell self-renewal and proliferation. Here some pathways are shown.

As per the previous study on Hedgehog signaling as well as its role in embryonic pattern formation is very clear, However, still it functions in adult tissue renewal, maintenance and relationship with cancer development are still elusive. The previous study revealed that loss of Smoothened (SMO)—an essential component of Hh pathway—results in impaired hematopoietic stem cell renewal and decreases induction of chronic myelogenous leukemia (CML) by oncoprotein BCR-ABL1 (Zhao et al. 2009).

Further, as per the study on Wnt Signaling pathway also showed their role in the regulation of both self-renewal and oncogenesis in different organs. The canonical Wnt cascade has emerged as a critical regulator of stem cells. In many tissues, activation of Wnt signaling has also been associated with cancer (Reya and Clevers 2005). Intestinal cancer is initiated by Wnt—pathway—activating mutations in genes such as adenomatous polyposis coli (APC). As in most cancers, the cell of origin has remained elusive (Barker et al. 2009).

Many other pathways are also associated with oncogenes like Notch, STAT3, and NFkB signaling pathway. For examples previous studies on Notch activation through ligand Jagged-1 which consistently increased the primitive progenitor activity observed in Invitro and Invivo HSCs cultures. Which support its role in HSC self-renewal or maintenance of multipolarity (Zhang and Grivennikov 2013). As per the previous studies, the role of STAT3 signaling in tumor environment is associated with increased T cell accumulation and higher expression of cytokines like CCL5 and CXCL10 inside the tumor environment, thus, the activation of STAT3 signaling is associated with another oncogenic pathway through reduction of immune cell recruitment in tumor microenvironment (Ivanov and Honda 2012). While studies on NFκB signaling suggested its role in tumor progression and growth rather interaction with the immune system, however, the activation of NF-κB via increased production of tumor-derived chemokines has a positive effect on T cell infiltration. However, this effect majorly depends on the cellular context of the tumor and type of immune infiltrates (Zhang and Grivennikov 2013).

4.4 CSC Niche and Its Contribution to Tumor Therapy

Normal stem cells reside in a "stem cell niche" that maintains them in a stem-like state. Based on the recent studies mostly CSCs also stayed on niche, called CSCs niche, which is associated with its self-renewal and differentiation, provides their blood supply and protects them against the immune system. In normal circumstances, stem cell niche provides only instantaneous proliferating signal which is necessary for cell regeneration. While in cancer, CSCs resist against the control of the niche and start infinite proliferation. Deeper understanding of the niche may be important to understand the CSCs properties (Borovski 2011; Flesken-Nikitin et al. 2013; Dionne et al. 2015). Hence specific therapeutic treatments targeting CSC niche are developed, and these methods are also combined with conventional therapies like radio- and chemotherapies, thus helping improve cancer treatment.

4.5 Immune Response and Immunotherapy for Cancer Stem Cells

4.5.1 Cancer Immunosurveillance and Immunoediting for CSCs

Immune system cancer cell interactions have been a subject of much controversy throughout the twentieth century. In 1909, Paul Ehrlich proposed that immune system suppressed oncogenesis in long-lived mammals (Ehrlich 1909). Five decades later, F. MacFarlane Burnet proposed the term of "immunosurveillance" which is used to define a process in immune system, especially the activities of T cells during the recognition and elimination of transformed cells (Burnet 1957). Although this hypothesis received considerable challenges over the next several decades (Stutman 1974), results accumulated during the 1990s, which are promoted by mouse genetics and monoclonal antibody technologies. These results finally validated the cancer immunosurveillance concept (Smyth et al. 2001). Recognition of behavioral traits by the innate and adaptive immune cells arising due to malignant transformation lies at the heart of cancer immunosurveillance (Vesely et al. 2011). In knockout mouse models, interferon-γ (IFNγ) (Shankaran et al. 2001; Dighe et al. 1994), interleukin 12 (IL-12) (Zitvogel and Kroemer 2014), perforin (van den Broek et al. 1996), and tumor necrosis factor related apoptosis-inducing ligand (TRAIL) (Frew et al. 2005) represent the key molecules associated with tumor immune-surveillance. The lack of any of these molecules results in augmented spontaneous tumorigenesis (Bruttel and Wischhusen 2014). Hence, the ability to evade immune destruction was thus recognized as an additional hallmark of cancer (Hainaut and Plymoth 2013; Hanahan and Weinberg 2011).

Over the last decade and a half, there has been a resurgence of interest in cancer immunosurveillance. A novel concept of "cancer immunoediting," focusing on the idea that immunity manifests both a host-protective and a tumor-promoting effect on developing cancers, has evolved. Schreiber launched a new theory termed 3E theory (Dunn et al. 2004), which refers to Elimination, Equilibrium, as well as Escape. Here the proposed function of CSCs in "3E" is shown.

Elimination: Accumulation of oncogenic mutations in a cell over time causes malignancy. Such mutations in a differentiated cell result in upregulation of activatory NK cell receptor ligands and thus present tumor-specific peptides via MHC class Ia molecules. These cells can thus be identified by the immune system and eliminated, resulting in regeneration of non-cancerous cells.

Equilibrium: In contrast to other cells, stem cells (SC) live longer and also express a multitude of immunosuppressive factors. Hence, they might easily accumulate oncogenic mutations without the risk of being cleared by the immune system. However, CSCs are initially confined to stem cell niches and limited to asymmetric divisions. Differentiated daughter cells inherit all the malignant mutations; however, they are more immunogenic and thus can be eliminated by the adaptive immune system. Hence, an equilibrium may be established.

Escape: Defects in tumor immunosurveillance, immunoediting, or the expansion of CSCs may contribute to evasion of oncogenic cells from clearance. Additionally, cell aging, immunosuppressive therapies or diseases or other factors may also abate the ability of the immune system to constrain tumor cell growth and progression. Immunoediting describes the evolutionary adaptation of individual tumor subclones to the selection pressure exerted by the immune system. It will finally result the expansion of subclone with less immunogenic or more immunosuppressive traits.

4.5.2 Immunotherapy for Cancer Stem Cells

Immunotherapy—a treatment based on the use of one's own immune system to fight against diseases—is quickly becoming a radical method for treatment of some cancers—along with surgery, radiation, and chemotherapy. Immunotherapy may involve a very generic or sometimes a considerable specific mode of activation. Immunotherapy of cancer currently includes the use of monoclonal antibodies, immune checkpoint inhibitors, cancer vaccines, and non-specific immunotherapies.

Targeting of CSCs with monoclonal antibodies represents a novel strategy against cancer. Monoclonal antibodies specific for CSC proteins have demonstrated significant efficacy in reducing cancer in mice. Some have reported antitumor activity in clinical settings as well (Luis Tume et al. 2016).

The CSC vaccination is a substance entering the body to induce an immune response to diseases like cancers. Studies on vaccination against antigen ALDH1A1+ of CSCs have achieved great progress (Miyata et al. 2017). Vaccination of CSCs has been proved to be valid in killing CSCs and resisting against tumor recurrence (Duarte et al. 2013; Ning et al. 2012a; Ning et al. 2012b). Kuai et al. designed a novel nanotechnology for tumor vaccination: nanodiscs that can be loaded with tumor antigens and carried into tumor-bearing mice are able to drive a strong immune response against tumors (Kuai et al. 2017).

Depending on studies of immunotherapy and other kinds of treatments, the combination of different therapies could successfully fight cancer (Swaminathan et al. 2013).

4.6 Prospective

It is likely that future immunotherapies could implement a combinational approach for combating cancer cells. It could contain both tumor cell targeted way and CSCs targeted specific method as well as antitumor immune enhancement therapies. Only by developing effective immune methods against both CSCs and tumor cells, a better future of long-term survival of cancer patients could be achieved.

References

Ahmed N, Abubaker K, Findlay J, Quinn M (2013) Cancerous ovarian stem cells: obscure targets for therapy but relevant to chemoresistance. J Cell Biochem 114(1):21–34. https://doi.org/10.1002/jcb.24317

Al-Dhfyan A, Alhoshani A, Korashy HM (2017) Aryl hydrocarbon receptor/cytochrome P450 1A1 pathway mediates breast cancer stem cells expansion through PTEN inhibition and beta-catenin and Akt activation. Mol Cancer 16:69. https://doi.org/10.1186/s12943-017-0636-5

Barker N, van Es JH, Kuipers J, Kujala P, van den Born M, Cozijnsen M et al (2007) Identification of stem cells in small intestine and colon by marker gene Lgr5. Nature 449(7165):1003–U1001. https://doi.org/10.1038/nature06196

Barker N, Ridgway RA, van Es JH, van de Wetering M, Begthel H, van den Born M et al (2009) Crypt stem cells as the cells-of-origin of intestinal cancer. Nature 457(7229):608–U119. https://doi.org/10.1038/nature07602

Barr MP, Singh S, Foley E, He Y, Young V, Ryan R et al (2017) XRCC6BP1: a key player in cisplatin resistance and lung cancer stem cells. Lung Cancer 103:S1–S2

Bhatlekar S, Viswanathan V, Fields JZ, Boman BM (2017) Overexpression of HOXA4 and HOXA9 genes promotes self-renewal and contributes to colon cancer stem cell overpopulation. J Cell Physiol 233:727–735. https://doi.org/10.1002/jcp.25981

Borovski T, Melo FDE, Vermeulen L, Medema JP (2011) Cancer stem cell niche: the place to be. Cancer Res 71(3):634–639. https://doi.org/10.1158/0008-5472.Can-10-3220

Bruttel VS, Wischhusen J (2014) Cancer stem cell immunology: key to understanding tumorigenesis and tumor immune escape? Front Immunol 5:360. https://doi.org/10.3389/fimmu.2014.00360

Burnet M (1957) Cancer—a biological approach. 3. Viruses associated with neoplastic conditions. BMJ 1:841–846

Ceder J, Elgqvist J (2017) Targeting prostate cancer stem cells with alpha-particle therapy. Front Oncolgy 6:273. https://doi.org/10.3389/fonc.2016.00273

Dighe AS, Richards E, Old LJ, Schreiber RD (1994) Enhanced in-vivo growth and resistance to rejection of tumor-cells expressing dominant-negative Ifn-gamma receptors. Immunity 1(6):447–456. https://doi.org/10.1016/1074-7613(94)90087-6

Dionne LK, Driver ER, Wang XJ (2015) Head and neck cancer stem cells: from identification to tumor immune network. J Dent Res 94(11):1524–1531. https://doi.org/10.1177/0022034515599766

Duarte S, Momier D, Baqué P, Casanova V, Loubat A, Samson M, Guigonis JM, Staccini P, Saint-Paul MC, De Lima MP, Carle GF, Pierrefite-Carle V (2013) Preventive cancer stem cell-based vaccination reduces liver metastasis development in a rat colon carcinoma syngeneic model. Stem Cells 31:423–432

Dunn GP, Old LJ, Schreiber RD (2004) The three Es of cancer immunoediting. Annu Rev Immunol 22:329–360. https://doi.org/10.1146/annurev.immunol.22.012703.104803

Eaves CJ (2008) Cancer stem cells: here, there, everywhere? Nature 456(7222):581–582. https://doi.org/10.1038/456581a

Ehrlich P (1909) Ueber den jetzigen Stand der Karzinomforschung. Vortrag gehalten vor den Studenten der Amsterdamer Universitaet, Vereinigung fuer wissenschaftliche Arbeit 1 June 1908. Printed in: P. Ehrlich[J]. Beitraege zur Experimentellen Pathologie und Chemotherapie, Akademische Verlagsgesellschaft, Leipzig. 118–164

Flesken-Nikitin A, Hwang CI, Cheng CY, Michurina TV, Enikolopov G, Nikitin AY (2013) Ovarian surface epithelium at the junction area contains a cancer-prone stem cell niche. Nature 495(7440):241–245. https://doi.org/10.1038/nature11979

Frew AJ, Zerafa N, Westwood JA, Cretney E, Mitchell S, Waring P et al (2005) TRAIL-mediated immunosurveillance of malignancy. Tissue Antigens 66(5):420–420

Hainaut P, Plymoth A (2013) Targeting the hallmarks of cancer: towards a rational approach to next-generation cancer therapy. Curr Opin Oncol 25(1):50–51. https://doi.org/10.1097/CCO.0b013e32835b651e

Hanahan D, Weinberg RA (2011) Hallmarks of cancer: the next generation. Cell 144(5):646–674. https://doi.org/10.1016/j.cell.2011.02.013

Holah NS, Aiad HAES, Asaad NY, Elkhouly EA, Lasheen AG (2017) Evaluation of the role of ALDH1 as cancer stem cell marker in colorectal carcinoma: an immunohistochemical study. J Clin Diagn Res 11(1):Ec17–Ec23. https://doi.org/10.7860/Jcdr/2017/22671.9291

Ivanov II, Honda K (2012) Intestinal commensal microbes as immune modulators. Cell Host Microbe 12(4):496–508. https://doi.org/10.1016/j.chom.2012.09.009

Ke J, Wu X, Wu X, He X, Lian L, Zou Y et al (2012) A subpopulation of CD24(+) cells in colon cancer cell lines possess stem cell characteristics. Neoplasma 59(3):282–288. https://doi.org/10.4149/neo_2012_036

Kuai R, Ochyl LJ, Bahjat KS, Schwendeman A, Moon JJ (2017) Designer vaccine nanodiscs for personalized cancer immunotherapy. Nat Mater 16(4):489–496. https://doi.org/10.1038/Nmat4822

Lapidot T, Sirard C, Vormoor J, Murdoch B, Hoang T, Caceres-Cortes J et al (1994) A cell initiating human acute myeloid leukaemia after transplantation into SCID mice. Nature 367(6464):645–648. https://doi.org/10.1038/367645a0

Luis Tume KP, Ubidia-Incio R, Moya J (2016) CD133 in breast cancer cells and in breast cancer stem cells as another target for immunotherapy. Gaceta Mexicana de Oncología 15(1):22–30

Lupia M, Cavallaro U (2017) Ovarian cancer stem cells: still an elusive entity? Mol Cancer 16:64. https://doi.org/10.1186/s12943-017-0638-3

Maccalli C, Volonte A, Cimminiello C, Parmiani G (2014) Immunology of cancer stem cells in solid tumours. A review. Eur J Cancer 50(3):649–655. https://doi.org/10.1016/j.ejca.2013.11.014

Margaritescu C, Pirici D, Cherciu I, Barbalan A, Cartana T, Saftoiu A (2014) CD133/CD166/Ki-67 triple immunofluorescence assessment for putative cancer stem cells in colon carcinoma. J Gastrointestin Liver Dis 23(2):161–170

Medema JP (2013) Cancer stem cells: the challenges ahead. Nat Cell Biol 15(4):338–344. https://doi.org/10.1038/ncb2717

Melo FDE, Kurtova AV, Harnoss JM, Kljavin N, Hoeck JD, Hung J et al (2017) A distinct role for Lgr5(+) stem cells in primary and metastatic colon cancer. Nature 543(7647):676–680. https://doi.org/10.1038/nature21713

Miyata T, Oyama T, Yoshimatsu T, Higa H, Kawano D, Sekimura A et al (2017) The clinical significance of cancer stem cell markers ALDH1A1 and CD133 in lung adenocarcinoma. Anticancer Res 37(5):2541–2547. https://doi.org/10.21873/anticanres.11597

Ning N, Pan Q, Zheng F, Teitz-Tennenbaum S, Egenti M, Ginestier C et al (2012a) Cancer stem cell vaccination confers significant anti-tumor immunity by selectively targeting cancer stem cells. J Clin Immunol 32(2):358–358

Ning N, Pan Q, Zheng F, Teitz-Tennenbaum S, Egenti M, Yet J et al (2012b) Cancer stem cell vaccination confers significant antitumor immunity. Cancer Res 72(7):1853–1864. https://doi.org/10.1158/0008-5472.CAN-11-1400

Picot T, Aanei CM, Fayard A, Flandrin-Gresta P, Tondeur S, Gouttenoire M et al (2017) Expression of embryonic stem cell markers in acute myeloid leukemia. Tumour Biol 39(7):1010428317716629. https://doi.org/10.1177/1010428317716629

Reya T, Clevers H (2005) Wnt signalling in stem cells and cancer. Nature 434(7035):843–850. https://doi.org/10.1038/nature03319

Shankaran V, Ikeda H, Bruce AT, White JM, Swanson PE, Old LJ, Schreiber RD (2001) IFN gamma and lymphocytes prevent primary tumour development and shape tumour immunogenicity. Nature 410(6832):1107–1111. https://doi.org/10.1038/35074122

Singh K, Jain M, Agarwal K, Pathania O (2017) Cancer stem cell marker, CD44 expression in Indian breast carcinoma patients and its association with the molecular subtypes of breast carcinoma. Breast 32:S105–S105

Smyth MJ, Godfrey DI, Trapani JA (2001) A fresh look at tumor immunosurveillance and immunotherapy. Nat Immunol 2(4):293–299. https://doi.org/10.1038/86297

Stigbrand T (2017) Lgr5-positive cells are cancer stem cells in skin squamous cell carcinoma (Retraction of Vol 35, Pg 11605, 2014). Tumor Biol

Stutman O (1974) Tumor development after 3-methylcholanthrene in immunologically deficient athymic nude mice. Science 183(4124):534–536. https://doi.org/10.1126/science.183.4124.534

Sullivan KE, Rojas K, Cerione RA, Nakano I, Wilson KF (2017) The stem cell/cancer stem cell marker ALDH1A3 regulates the expression of the survival factor tissue transglutaminase, in mesenchymal glioma stem cells. Oncotarget 8(14):22325–22343. https://doi.org/10.18632/oncotarget.16479

Swaminathan SK, Roger E, Toti U, Niu L, Ohlfest JR, Panyam J (2013) CD133-targeted paclitaxel delivery inhibits local tumor recurrence in a mouse model of breast cancer. J Control Release 171:280–287

Tachezy M, Zander H, Wolters-Eisfeld G, Muller J, Wicklein D, Gebauer F et al (2014) Activated leukocyte cell adhesion molecule (CD166): an "inert" cancer stem cell marker for non-small cell lung cancer? Stem Cells 32(6):1429–1436. https://doi.org/10.1002/stem.1665

van den Broek MF, Kagi D, Ossendorp F, Toes R, Vamvakas S, Lutz WK et al (1996) Decreased tumor surveillance in perforin-deficient mice. J Exp Med 184(5):1781–1790. https://doi.org/10.1084/jem.184.5.1781

Vesely MD, Kershaw MH, Schreiber RD, Smyth MJ (2011) Natural innate and adaptive immunity to cancer. Annu Rev Immunol 29(29):235–271. https://doi.org/10.1146/annurev-immunol-031210-101324

Wang J, Wu YY, Gao W, Li F, Bo YF, Zhu MX et al (2017a) Identification and characterization of CD133(+) CD44(+) cancer stem cells from human laryngeal squamous cell carcinoma cell lines. J Cancer 8(3):497–506. https://doi.org/10.7150/jca.17444

Wang JH, Wang LW, Ho CT, Zhang KS, Liu Q, Zhao H (2017b) Garcinol from Garcinia indica downregulates cancer stem-like cell biomarker ALDH1A1 in nonsmall cell lung cancer A549 cells through DDIT3 activation. J Agric Food Chem 65(18):3675–3683. https://doi.org/10.1021/acs.jafc.7b00346

Wang W, Li Y, Liu N, Gao Y, Li L (2017c) MiR-23b controls ALDH1A1 expression in cervical cancer stem cells. BMC Cancer 17:292. https://doi.org/10.1186/s12885-017-3192-x

Wang Y, Zhang C, Zhu H, Tang J, Zhang S, Luo J, Sun X (2017d) CD90 positive cells exhibit aggressive radioresistance in esophageal squamous cell carcinoma. J Thorac Dis 9(3):610–620. https://doi.org/10.21037/jtd.2017.03.28

Xiao M, Yan M, Zhang J, Xu Q, Qi S, Wang X, Chen W (2017) Cancer stem-like cell related protein CD166 degrades through E3 ubiquitin ligase CHIP in head and neck cancer. Exp Cell Res 353(1):46–53. https://doi.org/10.1016/j.yexcr.2017.03.005

Yanai H, Atsumi N, Tanaka T, Nakamura N, Komai Y, Omachi T et al (2017) Intestinal cancer stem cells marked by Bmi1 or Lgr5 expression contribute to tumor propagation via clonal expansion. Sci Rep 7:41838. https://doi.org/10.1038/srep41838
Yang R, An LY, Miao QF, Li FM, Han Y, Wang HX et al (2016) Effective elimination of liver cancer stem-like cells by CD90 antibody targeted thermosensitive magnetoliposomes. Oncotarget 7(24):35894–35916. https://doi.org/10.18632/oncotarget.9116
Zage PE, Whittle SB, Shohet JM (2017) CD114: a new member of the neural crest-derived cancer stem cell marker family. J Cell Biochem 118(2):221–231. https://doi.org/10.1002/jcb.25656
Zhang WZ, Grivennikov SI (2013) Top Notch cancer stem cells by paracrine NF-kappa B signaling in breast cancer. Breast Cancer Res 15(5):316. https://doi.org/10.1186/bcr3565
Zhang L, Guo X, Zhang D, Fan Y, Qin L, Dong S, Zhang L (2017) Upregulated miR-132 in Lgr5+ gastric cancer stem cell-like cells contributes to cisplatin-resistance via SIRT1/CREB/ABCG2 signaling pathway. Mol Carcinog 56:2022–2034. https://doi.org/10.1002/mc.22656
Zhao C, Chen A, Jamieson CH, Fereshteh M, Abrahamsson A, Blum J et al (2009) Hedgehog signalling is essential for maintenance of cancer stem cells in myeloid leukaemia. Nature 458(7239):776–U117. https://doi.org/10.1038/nature07737
Zitvogel L, Kroemer G (2014) CD103(+) dendritic cells producing interleukin-12 in anticancer immunosurveillance. Cancer Cell 26(5):591–593. https://doi.org/10.1016/j.ccell.2014.10.008

Chapter 5
Human Hepatic Stem/Progenitor Cells in Cancer and Liver Disease

Daniela Fanni, Clara Gerosa, Federica Lai, and Gavino Faa

Abbreviations

CC	Cholangiocellular carcinoma
CD	Cluster differentiation
CXCL	Chemokine (C-X-C motif) ligand
CYP	Cytochrome oxidase system
HCC	Hepatocellular carcinoma
IL	Interleukin
JAG	Jagged
PBC	Primary biliary cholangitis
TNF	Tumor necrosis factor

5.1 Historical and Scientific Background

The documentation and the characterization of the human stem/progenitor cells of the liver are an interesting subject of the current scientific literature (Fanni et al. 2016a). The stem/progenitor cell is specifically characterized by cellular plasticity, skill to differentiate into multiple cellular populations, and the self-renewal ability (Wagers et al. 2002). The regenerative potential of the adult human liver after a damage is based on the aptitude of hepatic stem/progenitor cells to differentiate both into hepatocytic or cholangiocyte lineage (Spee et al. 2010; Hirose et al. 2009). Besides liver stem/progenitor cells may also differentiate into extrahepatic cell types, like intestinal, pancreatic (Strain and Crosby 2000; Thorgeirsson 1996), and insulin-producing cells, or may give rise to liver cancer (Liu et al. 2013).

D. Fanni (✉) · C. Gerosa · F. Lai · G. Faa
Division of Pathology, University of Cagliari, AOU Cagliari, Cagliari, Italy

P. V. Pham, A. El-Hashash (eds.), *Stem Cells for Cancer and Genetic Disease Treatment*, Stem Cells in Clinical Applications, https://doi.org/10.1007/978-3-319-98065-2_5

The presence of hepatic stem/progenitor cells was discovered in human liver (Sell 1990), in human liver disease (De Vos and Desmet 1992) and cirrhosis (Baumann et al. 1999). Liver carcinogenesis studies gave the major support for further studies (Sell 1990). Actually, the proliferation control defeat and the differentiation arrest precede the hepatocellular carcinoma (Sigal et al. 1992; Sell 1993) and the cholangiocarcinomas (Yamashita et al. 2008). Hepatic stem/progenitor cells were identified in hepatoblastoma (Ruck et al. 1997) as well. The identification of hepatic stem/progenitor cells was first claimed in experimental animal models (Wilson and Leduc 1958; Dunsford and Sell 1989), where prolonged and severe hepatic injury with acetylaminofluorene in rats fed with choline devoid diet was restored by pre-existing parenchymal cells derived from undifferentiated cholangiole cells (Strain and Crosby 2000; Sell et al. 1981). The liver sections from these rat models prove the presence of proliferating small periportal scant cytoplasm oval cells, that spread across the liver (Sell et al. 1981; Fausto 1990). Further studies then discovered other poisons able to induce oval cells in combination or alone (Strain and Crosby 2000). The markers expressed or upregulated by rat oval cells are cytokeratins 18/18, cytokeratins 19, cytokeratins 7 (Faa et al. 1998; Fanni et al. 2016a), CD90, OV6, α-fetoprotein (Alison et al. 2009; Oertel and Shafritz 2008), CD34 and c-kit (Strain and Crosby 2000).

5.2 The Liver Niches

Liver stem/progenitor cells hypothesis embodies a heterogeneous population through a range of morphological and immunohistochemical features extending from bile duct cells to hepatocytes. According to the niche hypothesis, the stem/progenitor cells reside within a not specific compartment including the surrounding matrix, cells, and signals required for proliferation and differentiation control (Kordes and Haussinger 2013). The Wnt, Notch, and Hedgehog pathways regulate the stem/progenitor cell quiescence and proliferation. Particularly Notch and Wnt signals manage cell fate by regulating the switch between cells division or differentiation (Crosnier et al. 2006). The niche consists of the interaction with cell types, cellular and extracellular microenvironment, signaling and adhesion molecules besides the niche contributes to self-renewal and regulates stem/progenitor cellmaintenance (Crosnier et al. 2006; Moore and Lemischka 2006). Thus, the niche is not just the location of stem/progenitor cells (Kordes and Haussinger 2013). The hepatic niche is located in the canals of Hering (Kuwahara et al. 2008), lined either by hepatocytes or small cholangiocytes, at the periphery of the immature portal tract and in the mesenchyme near the ductal plate (Fanni et al. 2016a). The periportal zone gives rise to the small new hepatocytes, maturing and differentiating endlessly on the path to the central vein ("streaming liver") (Kordes and Haussinger 2013). The hepatic niche is composed of numerous cell types: hepatocytes, cholangiocytes, hepatic stellate cells, Kupffer cells, Pit cells, myofibroblasts, endothelial cells, and immune cells. All cells interact and cross-talk with hepatic stem/progenitor cells (Alison et al. 2009).

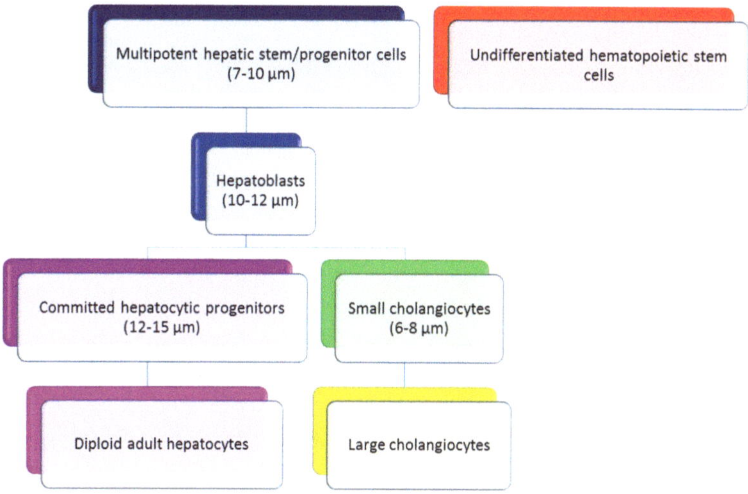

Fig. 5.1 Schematic representation of the hierarchical relationship of the niche three-cell lineage

Within the liver, three cell lineages should be taken into account (Fig. 5.1):

1. Hepatocellular
2. Ductal
3. Hematopoietic

5.3 Hepatocellular Lineage

Hepatocellular lineage includes the following cell subtypes:

1. Multipotent hepatic stem/progenitor cells (7–10 µm, great nucleus/cytoplasm ratio) (Stachelscheid et al. 2009; Zhou et al. 2007)
2. Hepatoblasts, larger (10–12 µm) with more cytoplasm (Turner et al. 2011)
3. Committed hepatocytic progenitors (12–15 µm) (Tan et al. 2002)
4. Diploid adult hepatocytes (small hepatocytes) (Kubota and Reid 2000)

Adult human hepatocytes contain cytokeratin 8 and 18 (Van Eyken et al. 1988). Differentiation of liver stem/progenitor cell toward hepatocyte evolves from the hepatoblast (cytokeratin 19 positive), through the committed hepatocytic intermediate cell with hepatobiliary phenotype (both cytokeratin 19 and cytokeratin 7 positive), to intermediate cell with hepatocytic phenotype (cytokeratin 19 negative and cytokeratin 7 positive) and finally to mature hepatocytes (cytokeratin 7 negative) (Desmet 2011).

From 9 weeks of gestational age, the portal vein branches are surrounded by a layer of cells, referred to as the ductal plate, showing a strong positivity reaction for

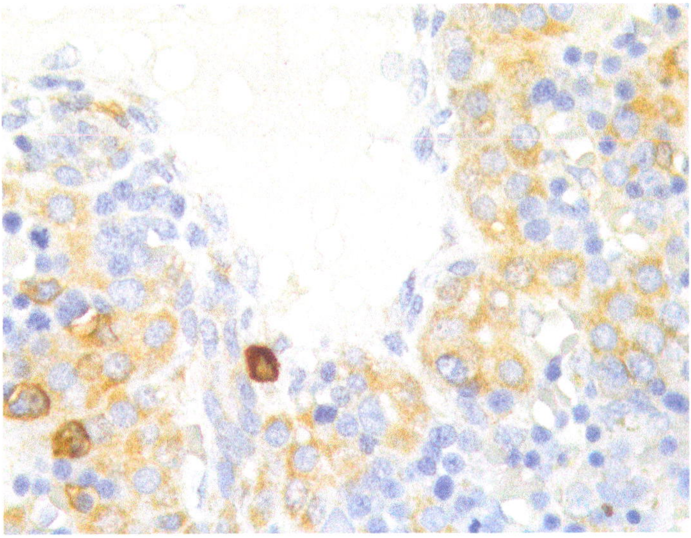

Fig. 5.2 Expression of c-kit in liver niche

CAM 5.2, KL-1, cytokeratin 8/18, cytokeratin 19 and negative for cytokeratin 7 until 20 weeks of gestation (Van Eyken et al. 1988).

Small undifferentiated oval cells (hepatoblasts) with scant cytoplasm are easily identified in the developing portal tracts. At the periphery of developing portal tracts in the ductal plate, these cells are easily identifiable by the expression of cytokeratin 7. Moreover, these cells are characterized by the expression of stem markers as SOX9 and c-kit (Fig. 5.2) (Fanni et al. 2016a).

The function specifically characterizing the hepatocyte is the ability to metabolize and detoxify through metabolic drug system based on the cytochrome P450 superfamily (CYP), made of monooxygenases critical for the metabolism of numerous endogenous and exogenous compounds (Fanni et al. 2014a). The human CYP3A family, being responsible for the metabolism of a large portion of toxins, carcinogens, drugs, and steroid hormones, represents the largest portion of human liver and includes CYP3A4, CYP3A5, CYP3A7, and CYP3A43(Zanger et al. 2005). CYP3A4 and CYP3A7, accounting for the majority of the oxidative metabolism of drugs, are the most representative of the CYP family human liver (Johnson et al. 2006; Bjorkman 2006; Kearns et al. 2003; de Wildt et al. 2002; Smits et al. 2012). The CYP3A4 and CYP3A7 immunohistochemical expression is strong and diffuse, being detectable in all acinar zone, from the hepatocytes of the periportal liver zone (zone 1 according to Rappaport) to surrounding the central (terminal) vein (zone 3 of Rappaport), in the fetal livers from 13 weeks of gestational ages (Fig. 5.3) and neonatal livers as well (Fanni et al. 2014b). Therefore, CYP3A4 and CYP3A7 are produced early by the hepatoblasts during liver development; on the other hand, they are not helpful to differentiate hepatoblasts from committed hepatocytic progenitors, diploid adult hepatocytes, and mature hepatocytes because of positive even in the adult and the neonatal liver.

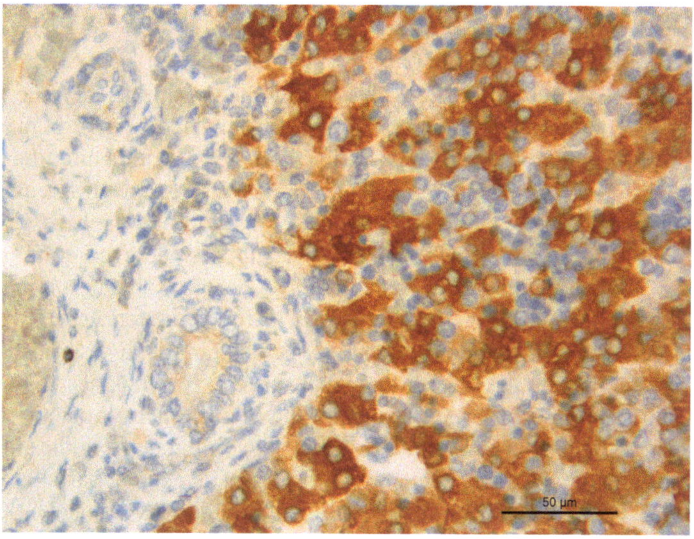

Fig. 5.3 CYP3A4 expression at 21 weeks of gestational age

5.4 Ductal Cell Lineage

Ductal lineage includes the following subtypes:

1. Multipotent hepatic stem/progenitor cells (7–10 µm, great nucleus/cytoplasm ratio) (Stachelscheid et al. 2009; Zhou et al. 2007)
2. Hepatoblasts, larger (10–12 µm) with more cytoplasm (Turner et al. 2011)
3. Small cholangiocytes (6–8 µm), with cuboidal shape, high nucleus/cytoplasm ratio, and small endoplasmic reticulum (Turner et al. 2011; Wang et al. 2010)
4. Large cholangiocytes, with columnar shape, small nucleus, conspicuous cytoplasm, an ample Golgi apparatus, and more rough endoplasmic reticulum (Turner et al. 2011; Ueno et al. 2003)

In contrast with adult human hepatocytes, bile duct cells also express either cytokeratin 7, cytokeratin 19, in addiction to cytokeratin 8 and cytokeratin 18 (Van Eyken et al. 1988). The development of the intrahepatic bile ducts in man can be studied by immunohistochemistry. From 9 weeks of gestational age, the ductal plate duplicates and forms tubular structures that are incorporated into the connective tissue surrounding the portal vein, resulting in the bile ducts (Van Eyken et al. 1988). Dedifferentiation of mature hepatocytes proceeding from hepatocyte (cytokeratin 7 negative) over the intermediate hepatobiliary cell (cytokeratin 7 positive) gives rise to the intermediate hepatobiliary cell (cytokeratin 7 positive and cytokeratin 19 positive) and hepatoblast (cytokeratin 19 positive and cytokeratin 7 negative) (Desmet 2011).

5.5 Hematopoietic Lineage

Hematopoiesis begins early in the yolk sac, then the liver turns out to be the major hematopoietic organ during embryonic and fetal development in humans. Liver hematopoietic activity begins during the first weeks of gestation after the colonization by undifferentiated hematopoietic stem cells, the regular differentiation and maturation, and the access in the circulatory system through the hepatic sinusoids (Fanni et al. 2018). Thus, while the liver niche supports the maturation of hepatoblasts in hepatocytes and cholangiocytes, the liver becomes able to support hematopoiesis as well constituting another niche for hematopoietic stem cells. Two types of cells demonstrate to play a role in supporting hematopoiesis: liver sinusoidal endothelial cells and hepatoblasts; both these cells produce chemokines (CXCL10 and CXCL12); hepatoblasts play a key role in the migration of hematopoietic stem cells. Hematopoietic stem cells C-kit+ and IL-7Rα+ likely migrate from the sinusoids to the hepatic parenchyma, making contact with sinusoidal endothelial cells and hepatoblasts. The hematopoietic stem cells differentiate into CD19+ and SLC+ pre-B cells (Kajikhina et al. 2015). Hepatoblasts and hepatic stellate cells produce erythropoietin and stem cell factor supporting hematopoiesis; hepatic stellate cells even produce CXCL12. Cells residing in the vascular walls (mesenchymal stromal cells) are part of the niche and have shown the ability to produce regulatory factors for hematopoietic stem cells (Payushina 2012). Great attention should be devoted to the morphology of hepatic hematopoiesis. The observation of sections obtained from the livers of fetuses at different weeks of gestation could already give important insights on the morphology of this hematopoietic stem cell niche, on how different lineages of blood cells develop during gestation, and on how the characteristics of hepatic hematopoiesis change depending on the gestational age. The hepatic hematopoiesis morphologically evolves as gestation progresses in human embryos and fetuses in four consecutive stages of liver hematopoiesis in humans (Fanni et al. 2018) (Fig. 5.4):

Stage I: hematopoiesis occupies 0–10% of the liver parenchyma, hematopoietic foci measure 9 × 11 μm, single hematopoietic immature cells are located in the sinusoids in contact with endothelial cells and among hepatoblasts, macrophages in the sinusoidal lumen.

Stage II: hematopoiesis occupies 30% of the parenchyma, hematopoietic foci are small, irregular, and spaced.

Stage III: hematopoiesis occupies ≥70% of the parenchyma, hematopoietic foci are ellipsoidal (35–70 μm) or cord (25 μm thick) shaped and in close contact between one another, macrophages at the center of the hematopoietic foci.

Stage IV: hematopoiesis occupies ≤50% of the parenchyma, solitary hematopoietic foci appear smaller (20 μm) and round shaped.

Hematopoiesis in the fetal liver ceases right after birth when the bone marrow becomes the only hematopoietic organ.

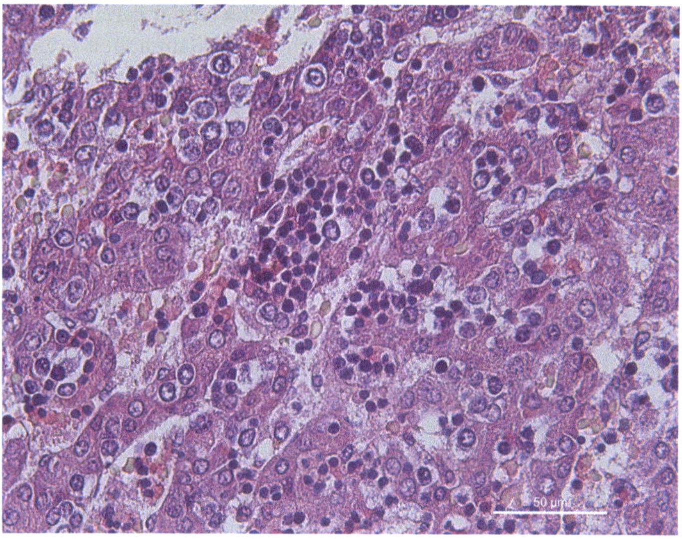

Fig. 5.4 Liver hematopoiesis at 14 weeks of gestational age

5.6 Carcinogenesis

The main primary liver malignancies are hepatocellular (HCC) and cholangiocellular carcinoma (CC), and mixed forms of both are described (Roskams et al. 2010; Van Haele and Roskams 2017). The range of immunohistochemical phenotypical traits of hepatocytes, cholangiocytes, and stem/progenitor cells seen in tumors is consistent with a stem/progenitor cell origin (Libbrecht et al. 2005). HCC is a major global disease that requires clinical, radiological, and pathological correlation for a proper diagnosis (European Association For The Study Of The Liver and European Organisation For Research And Treatment Of Cancer 2012). The histological differential diagnosis may be challenging because usually it is performed on liver biopsies and the focal lesion is characterized by marked morphological and immunohistochemical intratumoral variability (Senes et al. 2007). Liver stem/progenitor cells can differentiate into intrahepatic or extrahepatic cell types (Strain and Crosby 2000; Thorgeirsson 1996). Moreover, they may give rise to liver cancer (Liu et al. 2013), especially through the Wnt, Hedgehog, and Notch pathway disruption (Spee et al. 2010; Hirose et al. 2009). Liver carcinogenesis studies gave the major support for stem/progenitor cells knowledge (Sell 1990). During cancers development in transgenic mice oval cells, proliferation was observed (Bennoun et al. 1993; Schirmacher et al. 1991). Similarly, oval-like cells were described in human liver hepatocellular carcinomas and cholangiocarcinomas derived from hepatic stem/progenitor cells (Yamashita et al. 2008). Hepatic stem/progenitor cells were identified in hepatoblastoma (Ruck et al. 1997) as well. The developmental and differentiation arrest of hepatic stem/progenitor cells forego hepatocellular carcinoma (Sigal et al. 1992; Sell 1993). The oxidative stress and chronic inflammation are the carcinogenic risk factors in primary liver cancers since progenitor cells are activated in most chronic liver diseases. Thus, chronic hepatitis, toxic and metabolic diseases are the most important risk factors for the development of HCC, while chronic inflammatory biliary diseases are for the development of **CC** (Roskams et al. 2010; Van Haele and Roskams 2017). The tumors show their origin by phenotypical features, besides the liver tumors histopathologic classification is based on these features. HCC can express markers of stem/progenitor cells from biliary lineage: cytokeratins 7 (Fig. 5.5), cytokeratins 19, and OV6. The expression of cytokeratins 19 has been associated with a worse prognosis. The cells morphology ranges from very immature phenotype and intermediate phenotypes in between stem/progenitor cells and hepatocytes (Uenishi et al. 2003). According to the cancer stem/progenitor cell theory, HCC consists of hierarchical cell populations with different morphological and immunophenotypical features and different growth and metastatic potential (Roskams et al. 2010). Sorafenib is considered the first effective systemic treatment for HCC, thus it is currently the standard-of-care for patients with advanced HCC (European Association For The Study Of The Liver and European Organisation For Research And Treatment Of Cancer 2012). Sorafenib is metabolized by cytochrome P450 (CYP450), particularly from the CYP3A4. Similarly, with fetal human liver, HCCs express either CYP3A4 and CYP3A7. Specifically, the immunoreactivity of CYP3A4 was more strong and widespread than that of CYP3A7. The CYP3A4 immunoreactivity was observed in surrounding hepatocytes as well. Since sorafenib is metabolized by CYP3A4, but not by CYP3A7, an

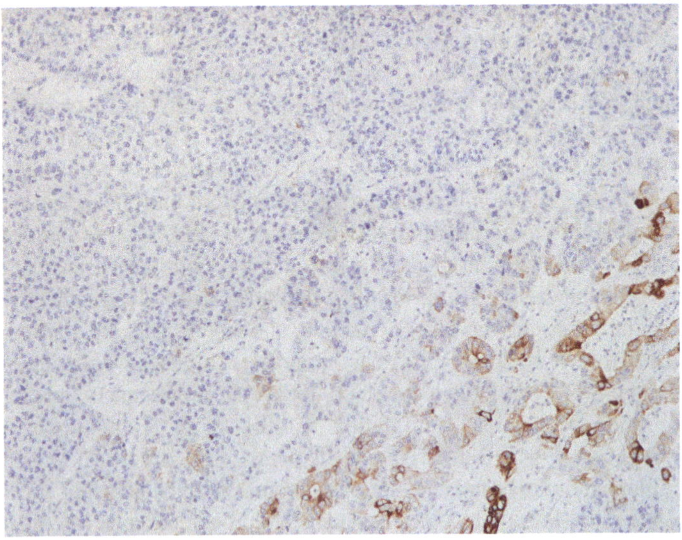

Fig. 5.5 Expression of cytokeratin 7 in HCC

overexpression of CYP3A4 may lead to an increase in the degradation of the drug and then to clinical ineffectiveness. These results might implicate the necessity of an individualized approach in the treatment of HCC as positivity to CYP3A4 in HCC liver samples might predict a scarce response to sorafenib (Fanni et al. 2016b). The identification of liver malignancy stem/progenitor cell markers and the understanding of their mechanisms of activation and differentiation can become the ideal target of new therapies (Roskams et al. 2010).

5.7 Liver Disease

The presence of small hepatic stem/progenitor cells was discovered in human liver disease (De Vos and Desmet 1992) and cirrhosis (Baumann et al. 1999). In acute and chronic hepatitis, the stem/progenitor cells activation correlates with the severity of the disease. The localization and severity of the inflammatory infiltrate in chronic viral hepatitis relates with the localization and the activation of stem/progenitor cells (Libbrecht et al. 2000). In acute hepatitis, activated stem/progenitor cells predominantly proliferate rather than differentiate. The differentiation toward hepatocytes delays after 1 week affecting the outcome and the prognosis (Katoonizadeh et al. 2006). In alcoholic steatohepatitis and primary biliary cirrhosis, stem/progenitor cells, within the liver niche, are capable of producing and responding to the ligands required for cells growth (Jung et al. 2007). The cytokines produced by the inflammatory infiltrate function as a chemotactic and growth factor for stem/progenitor cells. TNF super family, IL-6 family, lymphotoxin-b, interferon gamma, and TNF-a play an important role in stem/progenitor cells activation (Jakubowski et al. 2005; Streetz et al. 2003). SDF-1 is upregulated during human chronic liver disease, binds to the chemokine receptor 4 and plays a role in organogenesis cell proliferation, and trafficking (Terada et al. 2003). The Wnt and Notch signaling pathways play a key role in the stem cell proliferation regulation and differentiation toward committed lineages and are important during embryonic liver development (Kodama et al. 2004). The Wnt pathway works more in acute hepatitis then in HCV hepatitis virus or PBC (Crosnier et al. 2006). The role of the Notch pathway in bile duct development first became apparent in the Alagille syndrome, where the mutations of the Jagged1 gene dues the bile duct paucity. Notch 2 is responsible for the morphogenesis of bile duct structures (Geisler et al. 2008). Jagged1 and Notch 2 and 4 are more expressed in hepatic progenitor cells in biliary diseases than in hepatocytic diseases. Inhibition of the Notch pathway is necessary for hepatocyte differentiation (Roskams et al. 2010).The values of JAG1 increase in PBC in agreement with the expression of members of the Notch pathway by immature and mature cholangiocytes in vivo (Nijjar et al. 2002). The myofibroblasts positive for desmin and Notch 1 are present around stem/progenitor cells in PBC, indicating that they are responsible for the Notch cell–cell talking (Roskams et al. 2010). The liver niche gives the hope to new therapies for untreatable liver diseases (Alison et al. 2007).

References

Alison MR, Choong C, Lim S (2007) Application of liver stem cells for cell therapy. Semin Cell Dev Biol 18(6):819–826

Alison MR, Islam S, Lim S (2009) Stem cells in liver regeneration, fibrosis and cancer: the good, the bad and the ugly. J Pathol 217(2):282–298

Baumann U, Crosby HA, Ramani P, Kelly DA, Strain AJ (1999) Expression of the stem cell factor receptor c-kit in normal and diseased pediatric liver: identification of a human hepatic progenitor cell? Hepatology 30(1):112–117

Bennoun M, Rissel M, Engelhardt N, Guillouzo A, Briand P, Weber-Benarous A (1993) Oval cell proliferation in early stages of hepatocarcinogenesis in simian virus 40 large T transgenic mice. Am J Pathol 143(5):1326–1336

Bjorkman S (2006) Prediction of cytochrome p450-mediated hepatic drug clearance in neonates, infants and children: how accurate are available scaling methods? Clin Pharmacokinet 45(1):1–11

Crosnier C, Stamataki D, Lewis J (2006) Organizing cell renewal in the intestine: stem cells, signals and combinatorial control. Nat Rev Genet 7(5):349–359

De Vos R, Desmet V (1992) Ultrastructural characteristics of novel epithelial cell types identified in human pathologic liver specimens with chronic ductular reaction. Am J Pathol 140(6):1441–1450

de Wildt SN, Kearns GL, Hop WC, Murry DJ, Abdel-Rahman SM, van den Anker JN (2002) Pharmacokinetics and metabolism of oral midazolam in preterm infants. Br J Clin Pharmacol 53(4):390–392

Desmet VJ (2011) Ductal plates in hepatic ductular reactions. Hypothesis and implications. I. Types of ductular reaction reconsidered. Virchows Arch 458(3):251–259

Dunsford HA, Sell S (1989) Production of monoclonal antibodies to preneoplastic liver cell populations induced by chemical carcinogens in rats and to transplantable Morris hepatomas. Cancer Res 49(17):4887–4893

European Association For The Study Of The Liver, European Organisation For Research And Treatment Of Cancer (2012) EASL-EORTC clinical practice guidelines: management of hepatocellular carcinoma. J Hepatol 56(4):908–943

Faa G, Van Eyken P, Roskams T, Miyazaki H, Serreli S, Ambu R, Desmet VJ (1998) Expression of cytokeratin 20 in developing rat liver and in experimental models of ductular and oval cell proliferation. J. Hepatol 29(4):628–633.

Fanni D, Ambu R, Gerosa C, Nemolato S, Castagnola M, Van Eyken P, Faa G, Fanos V (2014a) Cytochrome P450 genetic polymorphism in neonatal drug metabolism: role and practical consequences towards a new drug culture in neonatology. Int J Immunopathol Pharmacol 27(1):5–13

Fanni D, Fanos V, Ambu R, Lai F, Gerosa C, Pampaloni P, Van Eyken P, Senes G, Castagnola M, Faa G (2014b) Overlapping between CYP3A4 and CYP3A7 expression in the fetal human liver during development. J Matern Fetal Neonatal Med:1–5

Fanni D, Gerosa C, Lai F, Van Eyken P, Faa G (2016a) Stem/progenitor cells in the developing human liver: morphological and immunohistochemical features. J Pediatr Neonatal Individ Med 5(2):e050205

Fanni D, Manchia M, Lai F, Gerosa C, Ambu R, Faa G (2016b) Immunohistochemical markers of CYP3A4 and CYP3A7: a new tool towards personalized pharmacotherapy of hepatocellular carcinoma. Eur J Histochem 60(2):2614

Fanni D, Angotzi F, Lai F, Gerosa C, Senes G, Fanos V, Faa G (2018) Four stages of hepatic hematopoiesis in human embryos and fetuses. J Matern Fetal Neonatal Med 31(6): 701–707

Fausto N (1990) Oval cells and liver carcinogenesis: an analysis of cell lineages in hepatic tumors using oncogene transfection techniques. Prog Clin Biol Res 331:325–334

Geisler F, Nagl F, Mazur PK, Lee M, Zimber-Strobl U, Strobl LJ, Radtke F, Schmid RM, Siveke JT (2008) Liver-specific inactivation of Notch2, but not Notch1, compromises intrahepatic bile duct development in mice. Hepatology 48(2):607–616

Hirose Y, Itoh T, Miyajima A (2009) Hedgehog signal activation coordinates proliferation and differentiation of fetal liver progenitor cells. Exp Cell Res 315(15):2648–2657

Jakubowski A, Ambrose C, Parr M, Lincecum JM, Wang MZ, Zheng TS, Browning B, Michaelson JS, Baetscher M, Wang B, Bissell DM, Burkly LC (2005) TWEAK induces liver progenitor cell proliferation. J Clin Invest 115(9):2330–2340

Johnson TN, Rostami-Hodjegan A, Tucker GT (2006) Prediction of the clearance of eleven drugs and associated variability in neonates, infants and children. Clin Pharmacokinet 45(9): 931–956

Jung Y, McCall SJ, Li YX, Diehl AM (2007) Bile ductules and stromal cells express hedgehog ligands and/or hedgehog target genes in primary biliary cirrhosis. Hepatology 45(5): 1091–1096

Kajikhina K, Tsuneto M, Melchers F (2015) Environments of hematopoiesis and B-lymphopoiesis in foetal liver. Clin Exp Rheumatol 33(92):S91–SS3

Katoonizadeh A, Nevens F, Verslype C, Pirenne J, Roskams T (2006) Liver regeneration in acute severe liver impairment: a clinicopathological correlation study. Liver Int 26(10): 1225–1233

Kearns GL, Robinson PK, Wilson JT, Wilson-Costello D, Knight GR, Ward RM, van den Anker JN (2003) Cisapride disposition in neonates and infants: in vivo reflection of cytochrome P450 3A4 ontogeny. Clin Pharmacol Ther 74(4):312–325

Kodama Y, Hijikata M, Kageyama R, Shimotohno K, Chiba T (2004) The role of notch signaling in the development of intrahepatic bile ducts. Gastroenterology 127(6):1775–1786

Kordes C, Haussinger D (2013) Hepatic stem cell niches. J Clin Invest 123(5):1874–1880

Kubota H, Reid LM (2000) Clonogenic hepatoblasts, common precursors for hepatocytic and biliary lineages, are lacking classical major histocompatibility complex class I antigen. Proc Natl Acad Sci U S A 97(22):12132–12137

Kuwahara R, Kofman AV, Landis CS, Swenson ES, Barendswaard E, Theise ND (2008) The hepatic stem cell niche: identification by label-retaining cell assay. Hepatology 47(6):1994–2002

Libbrecht L, Desmet V, Van Damme B, Roskams T (2000) Deep intralobular extension of human hepatic ‘progenitor cells’ correlates with parenchymal inflammation in chronic viral hepatitis: can ‘progenitor cells’ migrate? J Pathol 192(3):373–378

Libbrecht L, Desmet V, Roskams T (2005) Preneoplastic lesions in human hepatocarcinogenesis. Liver Int 25(1):16–27

Liu WH, Ren LN, Chen T, Liu LY, Tang LJ (2013) Stages based molecular mechanisms for generating cholangiocytes from liver stem/progenitor cells. World J Gastroenterol 19(41):7032–7041

Moore KA, Lemischka IR (2006) Stem cells and their niches. Science 311(5769):1880–1885

Nijjar SS, Wallace L, Crosby HA, Hubscher SG, Strain AJ (2002) Altered Notch ligand expression in human liver disease: further evidence for a role of the Notch signaling pathway in hepatic neovascularization and biliary ductular defects. Am J Pathol 160(5):1695–1703

Oertel M, Shafritz DA (2008) Stem cells, cell transplantation and liver repopulation. Biochim Biophys Acta 1782(2):61–74

Payushina OV (2012) Hematopoietic microenvironment in the fetal liver: roles of different cell populations. ISRN Cell Biol 2012:979480

Roskams T, Katoonizadeh A, Komuta M (2010) Hepatic progenitor cells: an update. Clin Liver Dis 14(4):705–718

Ruck P, Xiao JC, Pietsch T, Von Schweinitz D, Kaiserling E (1997) Hepatic stem-like cells in hepatoblastoma: expression of cytokeratin 7, albumin and oval cell associated antigens detected by OV-1 and OV-6. Histopathology 31(4):324–329

Schirmacher P, Held WA, Yang D, Biempica L, Rogler CE (1991) Selective amplification of periportal transitional cells precedes formation of hepatocellular carcinoma in SV40 large tag transgenic mice. Am J Pathol 139(1):231–241

Sell S (1990) Is there a liver stem cell? Cancer Res 50(13):3811–3815
Sell S (1993) Cellular origin of cancer: dedifferentiation or stem cell maturation arrest? Environ Health Perspect 101(Suppl 5):15–26
Sell S, Osborn K, Leffert HL (1981) Autoradiography of "oval cells" appearing rapidly in the livers of rats fed N-2-fluorenylacetamide in a choline devoid diet. Carcinogenesis 2(1):7–14
Senes G, Fanni D, Cois A, Uccheddu A, Faa G (2007) Intratumoral sampling variability in hepatocellular carcinoma: a case report. World J Gastroenterol 13(29):4019–4021
Sigal SH, Brill S, Fiorino AS, Reid LM (1992) The liver as a stem cell and lineage system. Am J Phys 263(2 Pt 1):G139–G148
Smits A, Kulo A, de Hoon JN, Allegaert K (2012) Pharmacokinetics of drugs in neonates: pattern recognition beyond compound specific observations. Curr Pharm Des 18(21):3119–3146
Spee B, Carpino G, Schotanus BA, Katoonizadeh A, Vander Borght S, Gaudio E, Roskams T (2010) Characterisation of the liver progenitor cell niche in liver diseases: potential involvement of Wnt and Notch signalling. Gut 59(2):247–257
Stachelscheid H, Urbaniak T, Ring A, Spengler B, Gerlach JC, Zeilinger K (2009) Isolation and characterization of adult human liver progenitors from ischemic liver tissue derived from therapeutic hepatectomies. Tissue Eng A 15(7):1633–1643
Strain AJ, Crosby HA (2000) Hepatic stem cells. Gut 46(6):743–745
Streetz KL, Tacke F, Leifeld L, Wustefeld T, Graw A, Klein C, Kamino K, Spengler U, Kreipe H, Kubicka S, Muller W, Manns MP, Trautwein C (2003) Interleukin 6/gp130-dependent pathways are protective during chronic liver diseases. Hepatology 38(1):218–229
Tan J, Hytiroglou P, Wieczorek R, Park YN, Thung SN, Arias B, Theise ND (2002) Immunohistochemical evidence for hepatic progenitor cells in liver diseases. Liver 22(5): 365–373
Terada R, Yamamoto K, Hakoda T, Shimada N, Okano N, Baba N, Ninomiya Y, Gershwin ME, Shiratori Y (2003) Stromal cell-derived factor-1 from biliary epithelial cells recruits CXCR4-positive cells: implications for inflammatory liver diseases. Lab Investig 83(5): 665–672
Thorgeirsson SS (1996) Hepatic stem cells in liver regeneration. FASEB J 10(11):1249–1256
Turner R, Lozoya O, Wang Y, Cardinale V, Gaudio E, Alpini G, Mendel G, Wauthier E, Barbier C, Alvaro D, Reid LM (2011) Human hepatic stem cell and maturational liver lineage biology. Hepatology 53(3):1035–1045
Uenishi T, Kubo S, Yamamoto T, Shuto T, Ogawa M, Tanaka H, Tanaka S, Kaneda K, Hirohashi K (2003) Cytokeratin 19 expression in hepatocellular carcinoma predicts early postoperative recurrence. Cancer Sci 94(10):851–857
Ueno Y, Alpini G, Yahagi K, Kanno N, Moritoki Y, Fukushima K, Glaser S, LeSage G, Shimosegawa T (2003) Evaluation of differential gene expression by microarray analysis in small and large cholangiocytes isolated from normal mice. Liver Int 23(6):449–459
Van Eyken P, Sciot R, Callea F, Van der Steen K, Moerman P, Desmet VJ (1988) The development of the intrahepatic bile ducts in man: a keratin-immunohistochemical study. Hepatology 8(6):1586–1595
Van Haele M, Roskams T (2017) Hepatic progenitor cells: an update. Gastroenterol Clin N Am 46(2):409–420
Wagers AJ, Christensen JL, Weissman IL (2002) Cell fate determination from stem cells. Gene Ther 9(10):606–612
Wang Y, Yao HL, Cui CB, Wauthier E, Barbier C, Costello MJ, Moss N, Yamauchi M, Sricholpech M, Gerber D, Loboa EG, Reid LM (2010) Paracrine signals from mesenchymal cell populations govern the expansion and differentiation of human hepatic stem cells to adult liver fates. Hepatology 52(4):1443–1454
Wilson JW, Leduc EH (1958) Role of cholangioles in restoration of the liver of the mouse after dietary injury. J Pathol Bacteriol 76(2):441–449
Yamashita T, Forgues M, Wang W, Kim JW, Ye Q, Jia H, Budhu A, Zanetti KA, Chen Y, Qin LX, Tang ZY, Wang XW (2008) EpCAM and alpha-fetoprotein expression defines novel prognostic subtypes of hepatocellular carcinoma. Cancer Res 68(5):1451–1461

Zanger UM, Klein K, Richter T, Toscano C, Zukunft J (2005) Impact of genetic polymorphism in relation to other factors on expression and function of human drug-metabolizing p450s. Toxicol Mech Methods 15(2):121–124

Zhou H, Rogler LE, Teperman L, Morgan G, Rogler CE (2007) Identification of hepatocytic and bile ductular cell lineages and candidate stem cells in bipolar ductular reactions in cirrhotic human liver. Hepatology 45(3):716–724

Chapter 6
Stem Cells for the Treatment of Ovarian Cancer

Kanwal Rehman and Muhammad Sajid Hamid Akash

Abbreviations

AACR	American Association for Cancer research
ALDH1A1	Aldehyde dehydrogenase-1A1
ATRA	All-trans-retinoic acid
CD	Cluster of differentiation
CSCs	Cancer stem cells
DM	Diabetes mellitus
EpCAM	Epithelial cell adhesion molecules
FAK	Focal adhesion kinase
SCs	Stem cells
STAT3	Signal transducer and activator of transcription 3
ZEB2	Zinc finger E-box-binding homeobox 2

6.1 Introduction

Ovarian cancer is one of the major causes of deaths in women suffering with gynecologic malignancies, which occurs at post-menopausal age (Ledermann et al. 2013) and ranks at 5th number after lung, breast, colorectal, and pancreatic cancers (Jemal et al. 2010) with severe impact on public health (Bast Jr. et al. 2009; Kusumbe and Bapat 2008). It has been documented that the estimated cases of ovarian cancer

K. Rehman
Institute of Pharmacy, Physiology and Pharmacology, University of Agriculture, Faisalabad, Punjab, Pakistan
e-mail: kanwal.akash@uaf.edu.pk

M. S. H. Akash (✉)
Department of Pharmaceutical Chemistry, Government College University Faisalabad, Faisalabad, Punjab, Pakistan
e-mail: sajidakash@gcuf.edu.pk

P. V. Pham, A. El-Hashash (eds.), *Stem Cells for Cancer and Genetic Disease Treatment*, Stem Cells in Clinical Applications, https://doi.org/10.1007/978-3-319-98065-2_6

in the USA was 22,240 along with estimated deaths of 14,030 (Jemal et al. 2009; Siegel et al. 2011). Ovarian cancer is the group of heterogeneous diseases, which are produced from different tissues of the female reproductive organ (Guddati 2012). The initial progression of ovarian cancer is unable to be detected, therefore it is frequently diagnosed at an advanced stage (Wikborn et al. 1996). The survival of the ovarian cancer patient depends on the disease lasting after cytoreductive surgery (Guddati 2012; Bristow et al. 2009). Malignancies that arise from the ovarian epithelium contain the vast majority of the cases of the ovarian cancer and in past, it has been thought that epithelium is the tissue of the origin for epithelial ovarian cancer (Murdoch and McDonnel 2002), but it has been experimentally proved that other than the epithelium tissues may also be involved in ovarian cancer (Dubeau 2008; Karst and Drapkin 2010). Even though, the notion is that repeated rupture of the epithelium during the ovulation may be responsible for various types of ovarian carcinogenesis (Ness and Cottreau 1999) and pathological specimens from surgeries have confirmed that epithelium in the fallopian tubes is responsible for the origin of different types of ovarian cancer (Carcangiu et al. 2004; Leeper et al. 2002).

It has been hypothesized (Auersperg et al. 2001; Flesken-Nikitin et al. 2013) that the origin of ovarian cancer stem cells (CSCs) results from the DNA damaging effects of the ovulation and wounding that affect the slow proliferating stem cells residing in the ovarian surface epithelium. It has been proposed that the presence of common genes on the surface of ovarian epithelium that are overexpressed in ovarian cancer, are involved in the development of ovarian cancer (Emmanuel et al. 2011), and develop a correlation between the ovulation and the development of ovarian cancer (Permuth-Wey and Sellers 2009).

In this chapter, we have briefly described the recent advances that have been made for the treatment of ovarian cancers and characterization of CSCs on the basis of their resistance to conventional chemotherapeutic agents. Moreover, we have also aimed to describe the most selected ovarian CSCs biomarkers that are directly related to the molecular signal transduction pathways and translation into the targeted treatment strategies.

6.2 Standard Treatment Strategies

Standard treatment strategy for ovarian cancer is the use of surgical resection of disease that follows the use of taxane- and platinum-based chemotherapy. This treatment strategy produces 80% of the partial response rate and 40–60% of the complete response rate (Society 2014) with 30% of 5-year survival rate (Kurman and Shih 2010), but the rate of recurrence of this disease is still 70% (Leitao Jr. and Chi 2009; Jordan 2004; Hope et al. 2003). Based on the initial response rate (80%), it indicates that initially, the majority of the ovarian cancers are chemosensitive, but the maximum chances of the recurrence (70%) of this disease indicate that some of the cancerous cells are not killed by chemotherapeutic agent and/or they regenerate after the exposure of chemotherapeutic agents (Farghaly 2014). Such type of cells

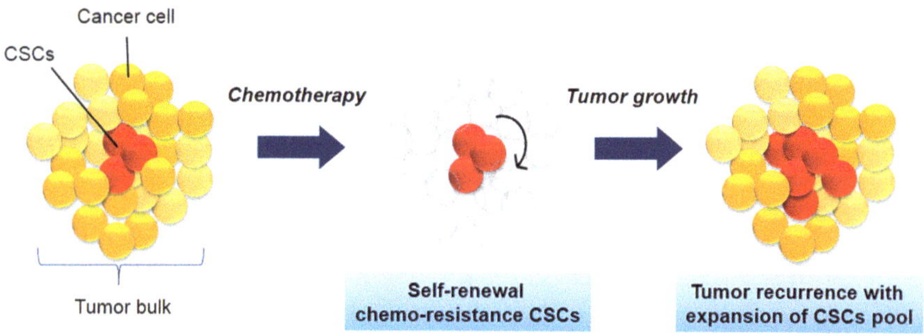

Fig. 6.1 Schematic representation of renewal and recurrence of cancer stem cells (CSCs). CSCs (red in color) and differentiate with in the tumor cells. After self-renewal, CSCs form non-tumorigenic cells (yellow). Adopted from Ref. Zakaria et al. (2017)

is known as CSCs. A schematic representation of self-renewal and recurrence of CSC has been described in Fig. 6.1 (Zakaria et al. 2017). According to the American Association for Cancer research (AACR) workshop, CSCs are the malignant cancer cells with a stem cell phenotype (Clarke et al. 2006). These are the malignant counterparts of the normal adult tissue stem cells (SCs) and are unable to maintain the SCs homeostasis due to dysregulation in signaling pathways (Tomao et al. 2013). It has been hypothesized that if the CSCs are responsible to cause relapse of disease, then they may be considered as an important therapeutic target for the elimination of ovarian cancer (Foster et al. 2013).

6.3 Biomarkers of Ovarian CSCs

Apparently, there are no known and/or specific biomarkers that may be used to define and detect ovarian CSCs, but certain protein-based biomarkers that have been identified by some other malignancies may be used for the detection of ovarian CSCs (Farghaly 2014). These protein-based biomarkers are involved in cell–matrix and cell–cell interactions. These protein-based biomarkers have the ability to affect the cellular growth, motility, and differentiation. In the following sub-sections, the important protein-based biomarkers have been described (Fig. 6.2) (Palmirotta et al. 2017).

6.3.1 CD44 Cells as a Biomarker of Ovarian CSCs

These cells have the ability to self-propagate and increase the chemoresistance against platinum- and taxane-based chemotherapy (Alvero et al. 2009). It has been found that these cells are enriched in ovarian cancer patient ascites and are thought to be the indicatives of CSCs phenotypes. Moreover, it has also been found that higher levels of CD44 expression along with other CSC biomarkers are correlated

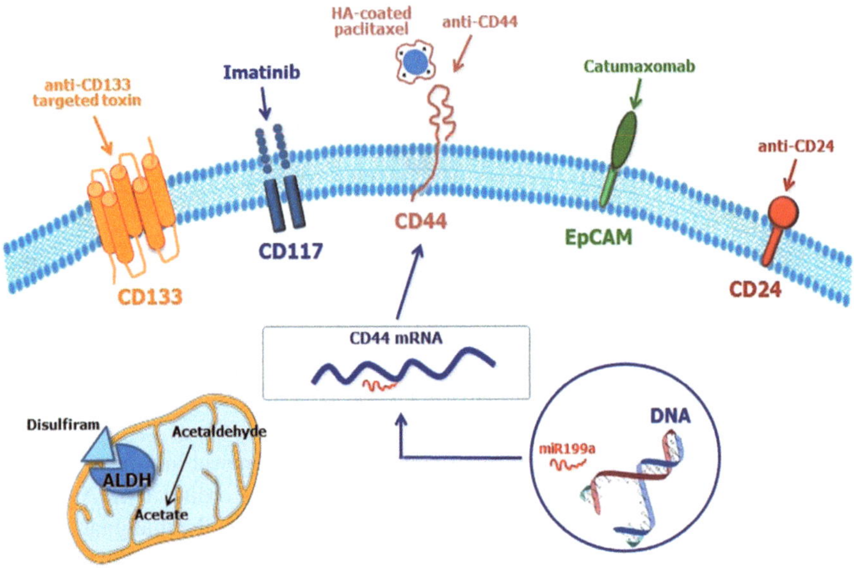

Fig. 6.2 Schematic representation of putative biomarkers for ovarian cancer stem cells CSCs). These putative biomarkers have the ability to counterattack the ovarian CSCs by providing the active sites for the mode of action of wide range of anticancer drugs. Adopted from Ref. Palmirotta et al. (2017)

with chemoresistance to paclitaxel and carboplatin (Alvero et al. 2009; Steffensen et al. 2011). Once isolated and xenografted, they give rise to the tumor with both $CD44^+$ and CD^- cells (Orian-Rousseau 2010). There are various ways to target the CD44 receptor, which are the binding hyaluronic acid and osteopontin. Osteopontin is a protein that is overexpressed in ovarian cancer and plays its role to activate the receptor tyrosine kinase (Ferrandina et al. 2008).

6.3.2 CD133 Cells as a Biomarker of Ovarian CSCs

CD133 cells also known as prominin-1 (Fargeas et al. 2003) are trans-membrane glycoproteins. These cells are highly expressed in human hematopoietic stem cells, progenitor cells, and mouse neuroepithelial cells (Foster et al. 2013). CD133 positive cells are higher in ovarian carcinoma than in normal ovarian tissues (Baba et al. 2009). These cells have the ability to recapitulate the original heterogeneous tumor and increase the chemoresistance tumor formation (Miettinen and Lasota 2005). These cells are used as biomarker for CSCs in hematopoietic tumors and variety of solid tumors including prostate, endometrial (Friel et al. 2008), neural (Liu et al. 2006; Singh et al. 2004), and ovarian (Baba et al. 2009; Curley et al. 2009; Ferrandina et al. 2009; Kryczek et al. 2012). Thought, the ability of $CD133^+$ cells to generate both $CD133^+$ and $CD133^-$ has been reported (Curley et al. 2009), $CD133^+$ cells generate large number of colonies and enhanced proliferative potential than those of $CD133^-$ cells (Ferrandina et al. 2008). Moreover, the percentages of $CD133^+$ epitopes expressing cells are significantly lower in normal ovaries with respect to those in ovarian tumors. Besides this, the percentages of these $CD133^+$ cells are also significantly lower in metastases than those in primary ovarian cancers.

6.3.3 CD177 Cells as a Biomarker of Ovarian CSCs

These are also known as stem cell growth factor receptors and highly expressed in the ovarian cancer cells (Luo et al. 2011), and highly tumorigenic in mouse models (Chau et al. 2013). They are involved in cell proliferation, differentiation, adhesion, and apoptosis (Kusumbe et al. 2009). CD117 cells activate Wnt/β-catenin pathway that plays a key role in the development of chemoresistance (Marchitti et al. 2008). In one study (Luo et al. 2011), it has been reported that the immunohistochemical analysis of 25 patients with serious ovarian cancers revealed that immunoreactivity of CD117 positive cells were detected in up to 40% of the patients and the expression of CD117 cells was strongly associated with the development of chemoresistance.

6.3.4 ALDH1A1 Cells as a Biomarker of Ovarian CSCs

Aldehyde dehydrogenase-1A1 (ALDH1A1) cells belong to a superfamily and contain 19 enzymes that protect the cells from carcinogenic aldehydes (Marchitti et al. 2008; Landen Jr. et al. 2010) and are associated with chemoresistance in ovarian carcinoma (Silva et al. 2011). These are intracellular enzymes that oxidize aldehydes and allow the conversion of retinol to retinoic acid by the involvement of differentiation pathways (Lawrenson and Gayther 2009; Moreb 2008; Shi et al. 2013). It has been found that the cells that are ++ for CD133 and ALDH1A1 are responsible to develop tumors in animal model (Shank et al. 2012), but metformin has the ability to decrease the formation of tumor spheres in patient tumors and population of $ALDH^+$ cells in ovarian cancer cell lines (Bellone et al. 2009). The detailed mechanism of action of metformin on ovarian CSCs has been schematically described in Fig. 6.3 (Kim et al. 2014). ALDH1A1 has been considered as a putative CSC biomarker that is associated with chemoresistance in ovarian CSCs (Tothill 2009). In one study, it has been found that the expression of ALDH1A1 and the enzymatic activity of ALDH were significantly higher in platinum- and taxane-resistant in ovarian cancer cell lines when compared with those of their corresponding drug sensitive parental cell lines (Landen Jr. et al. 2010). Similarly, in another study, it has also been reported that there was a distinct difference in the expression of ALDH1A1 between the normal epithelial tissue and epithelium tumor. Moreover, the higher expression of ALDH1A1 was also detected in serious ovarian cancer patients with poor clinical outcomes (Deng et al. 2010).

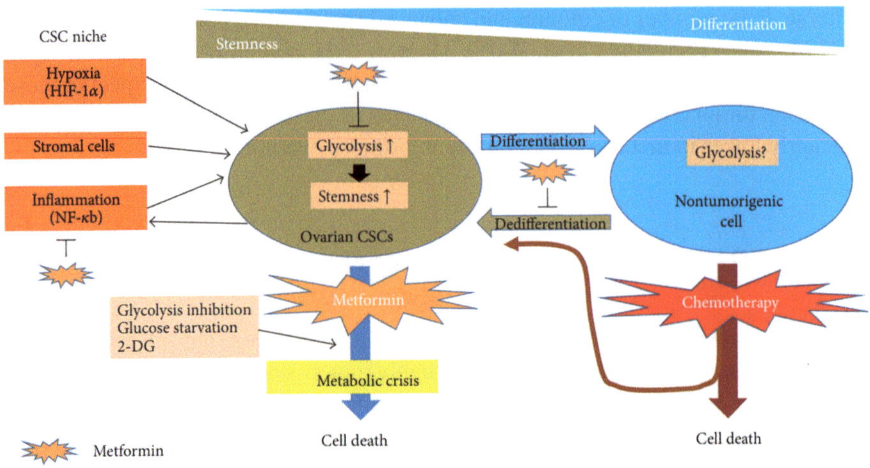

Fig. 6.3 Schematic representation of mode of action of metformin on ovarian cancer stem cells. A dynamic CSC model has been represented in the form of bidirectional interconversion between CSCs and differentiates non-CSCs. The characteristics and features of CSCs are mainly influenced by hypoxia, inflammation, and chemotherapy. Metformin exhibits its selective toxic effects on CSCs via involvement of metabolic stress that prevents the acquisition of stemness and dedifferentiation. In addition to this, metformin also exhibits its toxic effects by inducing the inflammatory mediators that are present in microenvironment. In response to metformin, ovarian CSCs are unable to cope with metformin-induced metabolic stress and ultimately cell death occurs. Adopted from Ref. Kim et al. (2014)

6.3.5 EpCAM Cells as a Biomarker of Ovarian CSCs

Epithelial cell adhesion molecules (EpCAM) are trans-membrane glycoproteins and also known as CD326. They are overexpressed in ovarian cancer and have the ability to show the oncogenic signaling properties. Their higher expression has been observed in metastatic ovarian tumors (Thiery et al. 2009). EpCAM cells are also involved in the transition of epithelial to mesenchymal cells that lead to metastasis (Choi et al. 2005). EpCAM is found to be associated with cytoplasmic staining in carcinoma tissues (Pauli et al. 2003).

6.3.6 CD24 Cells as a Biomarker of Ovarian CSCs

CD24 cells are glycosylphosphatidylinositol-linked cell surface proteins that are overexpressed in various types of solid tumors (Kristiansen et al. 2004) and have been considered as a putative CSC biomarker in ovarian cancer (Gao et al. 2010). Moreover, the cytoplasmic expression of CD24 has also been considered as CSC biomarker for the prediction of survival rates and recurrence of ovarian cancer

(Choi et al. 2005). They have shorter survival time in adenocarcinoma of ovary and their movement from cell membrane to cytoplasm in borderline ovarian tumors is associated with micro-invasion and omental implants (Moulla et al. 2013). It has been found that the transition of CD24 expression from membrane to cytoplasm is associated with a phenotype in borderline tumors (Scott et al. 2012). The attenuation and overexpression of CD24 regulate the phosphorylation of FAK and STAT3 by affecting the non-receptor tyrosine kinase activity (Tomao et al. 2013). It has been suggested that the contribution of CD24 expression is specific to the identification of CSC population in tumor specific and/or even patient specific (Foster et al. 2013).

6.4 Development of Chemoresistance to Conventional Therapies

Though the use of chemotherapeutic agent along with surgery is an effective treatment strategy to reduce the tumor mass, most of the patients acquire chemoresistance due to the residual of ovarian CSCs (Blagosklonny 2007; Clarke-Pearson 2009; Hanahan and Weinberg 2011). It has been described that even a small number of CSCs, if remain in situ after chemotherapy, may increase the incidences of recurrence of disease (Reya et al. 2001). It has been hypothesized that several pathways are involved in the mechanisms including the inactivation of pro-apoptotic effectors, activation of anti-apoptotic factors, and reinforcement of survival signals (Gimenez-Bonafe et al. 2009). Based on these mechanisms, the understanding of refractory response of CSCs to chemotherapy may attribute to drug effluxion, tumor dormancy, CSC dormancy, enrichment of CSCs during disease progression, glutathione system, and apoptosis. The description of these factors has been reviewed in detail by Tomao et al. (2013).

6.5 Therapeutic Strategies for Ovarian CSCs

The efficacy of any single drug against ovarian CSCs becomes limited due to their heterogeneity and it has been very challenging to the scientists to eliminate ovarian CSCs (Farghaly 2014). Moreover, the current first-line therapy for ovarian cancer is the use of combination of paclitaxel and carboplatin (Pignata et al. 2011), but it was found to be correlated with the increased incidences of chemoresistance (Zhang et al. 2008). Combination of more than one drug may be used to treat CSCs, but may increase the risk of toxicity as CSCs may also have the features of normal cells. This property of CSCs indicates that specialized tumor cells perform functions similar to that of normal stem cells. Even though, some therapeutic strategies have been used to treat ovarian CSCs that have been described in the following sub-sections.

6.5.1 Non-surface Biomarkers

Due to the remarkable success in the development of antibody-drug conjugates (Scott et al. 2012; Akash et al. 2016; Rehman and Akash 2017; Rehman et al. 2014), certain therapies have been developed that have the ability to target the biomarkers of ovarian CSCs and improve the oncogenic outcome and patient's survival (Burgos-Ojeda et al. 2012). It has been reported that many cell surface markers notably CD117, CD24, CD44, EpCAM, and CD133 have been used for the isolation of putative CSCs (Farghaly 2014). EpCAM is highly expressed in different types of tumors (Imrich et al. 2012), and its downregulation may lose the ability of cell–cell adhesion and promote the transition of epithelial mesenchymal cells (Thiery et al. 2009). Catumaxomab is monoclonal antibody against EpCAM and it is in phase III clinical trials for patients with malignant ascites (Seimetz et al. 2010). It is trifunctional antibody, which binds with three different cell types that are T-cells, tumor cells, and accessory cells (Sebastian et al. 2009). Moreover, Lin28 and Oct4 cells could also be used as targeted therapy for pluripotency in ovarian cancer (Peng et al. 2010).

6.5.2 Differentiation of Ovarian CSCs

Differentiation of ovarian CSCs results in the loss of their stemness property (Sell 2004). It has been reported that TWIST-1 plays a key role to eliminate ovarian CSCs by triggering the differentiation of epithelial ovarian cancer (Yin et al. 2013). It has been suggested that miRNAs can be used as therapeutic target for the treatment of cancers (Yu et al. 2012). Recently, it has been found that p53 activates miR-34a and miR-145 that promptly differentiate the human embryonic stem cells (Jain et al. 2012). Retinoic acid and its analogs are considered as the most important differentiation agents for ovarian CSCs. All-trans-retinoic acid (ATRA) has the ability to induce differentiation in head and neck squamous carcinoma CSCs by inhibiting the proliferation and Wnt/β-catenin pathway (Lim et al. 2012). It has also been found that ATRA induces the differentiation in acute promyelocytic leukemia, but limited in solid tumors (Soignet et al. 1998). Combination of retinoid compounds with carboplatin has been reported to have the ability to reduce the growth of ovarian CSCs (Whitworth et al. 2012). Similarly, linoleic acid has also shown to trigger the differentiation in ovarian cancer cells (Ruiz-Vela et al. 2011).

6.5.3 Niches of CSCs

These are the microenvironments that provide the favorable conditions to CSC for the growth, progression, and metastasis of CSCs (Ruiz-Vela et al. 2011). It has also been reported that tumorigenic ability of the ovarian tumor cells is dependent on the niches derived from human embryonic stem cells (Katz et al. 2009).

6.5.4 *MicroRNAs*

The expression profile of microRNAs is different between the normal cells and CSCs due to which they have the property of self-renewal and develop chemoresistance in ovarian CSCs (Xu et al. 2012). Therefore, they are involved in the biological processes of tumors by regulating the expression of genes at post-transcriptional level (Bartel 2009). It has been reported that miR-199a has the ability to prevent the tumorigenesis in xenograft model, while miR-200a reduces the migration of $CD133^{+}$ ovarian CSCs by inhibiting E-cadherin and ZEB2 (Wu et al. 2011). Similarly, mir214 that is highly expressed in ovarian CSCs has been shown to mediate the chemoresistance developed against cisplatin (Yang et al. 2008) via repressing p53-pathway (Xu et al. 2012).

6.6 Conclusion

Sufficient scientific literature provided in this chapter indicates that existence of CSCs in ovarian tumors is responsible to initiate the pathogenesis of ovarian cancer and tumors. Moreover, these CSCs also increase the chemoresistance in ovarian tumors along with the induction of more differentiated non-tumorigenic cells. To eliminate these CSCs, understanding the biology and functional role of these CSCs in ovarian tumors is very complex and recent advances in the identification of new ADCs to target these CSCs in ovarian cancers eradicate ovarian CSCs.

Conflict of Interest The authors declare that they do not have any conflict of interests for this chapter.

References

Akash MSH, Rehman K, Parveen A, Ibrahim M (2016) Antibody-drug conjugates as drug carrier systems for bioactive agents. Int J Polym Mater Polym Biomater 65(1):1–10

Alvero AB, Chen R, Fu HH, Montagna M, Schwartz PE, Rutherford T, Silasi DA, Steffensen KD, Waldstrom M, Visintin I (2009) and others. Molecular phenotyping of human ovarian cancer stem cells unravels the mechanisms for repair and chemoresistance. Cell Cycle 8(1):158–166

Auersperg N, Wong AS, Choi KC, Kang SK, Leung PC (2001) Ovarian surface epithelium: biology, endocrinology, and pathology. Endocr Rev 22(2):255–288

Baba T, Convery PA, Matsumura N, Whitaker RS, Kondoh E, Perry T, Huang Z, Bentley RC, Mori S, Fujii S et al (2009) Epigenetic regulation of CD133 and tumorigenicity of CD133+ ovarian cancer cells. Oncogene 28(2):209–218

Bartel DP (2009) MicroRNAs: target recognition and regulatory functions. Cell 136(2):215–233

Bast RC Jr, Hennessy B, Mills GB (2009) The biology of ovarian cancer: new opportunities for translation. Nat Rev Cancer 9(6):415–428

Bellone S, Siegel ER, Cocco E, Cargnelutti M, Silasi DA, Azodi M, Schwartz PE, Rutherford TJ, Pecorelli S, Santin AD (2009) Overexpression of epithelial cell adhesion molecule in primary,

metastatic, and recurrent/chemotherapy-resistant epithelial ovarian cancer: implications for epithelial cell adhesion molecule-specific immunotherapy. Int J Gynecol Cancer 19(5):860–866
Blagosklonny MV (2007) Cancer stem cell and cancer stemloids: from biology to therapy. Cancer Biol Ther 6(11):1684–1690
Bristow RE, Puri I, Chi DS (2009) Cytoreductive surgery for recurrent ovarian cancer: a meta-analysis. Gynecol Oncol 112(1):265–274
Burgos-Ojeda D, Rueda BR, Buckanovich RJ (2012) Ovarian cancer stem cell markers: prognostic and therapeutic implications. Cancer Lett 322(1):1–7
Carcangiu ML, Radice P, Manoukian S, Spatti G, Gobbo M, Pensotti V, Crucianelli R, Pasini B (2004) Atypical epithelial proliferation in fallopian tubes in prophylactic salpingo-oophorectomy specimens from BRCA1 and BRCA2 germline mutation carriers. Int J Gynecol Pathol 23(1):35–40
Chau WK, Ip CK, Mak AS, Lai HC, Wong AS (2013) c-Kit mediates chemoresistance and tumor-initiating capacity of ovarian cancer cells through activation of Wnt/beta-catenin-ATP-binding cassette G2 signaling. Oncogene 32(22):2767–2781
Choi YL, Kim SH, Shin YK, Hong YC, Lee SJ, Kang SY, Ahn G (2005) Cytoplasmic CD24 expression in advanced ovarian serous borderline tumors. Gynecol Oncol 97(2):379–386
Clarke MF, Dick JE, Dirks PB, Eaves CJ, Jamieson CH, Jones DL, Visvader J, Weissman IL, Wahl GM (2006) Cancer stem cells—perspectives on current status and future directions: AACR workshop on cancer stem cells. Cancer Res 66(19):9339–9344
Clarke-Pearson DL (2009) Clinical practice. Screening for ovarian cancer. N Engl J Med 361(2):170–177
Curley MD, Therrien VA, Cummings CL, Sergent PA, Koulouris CR, Friel AM, Roberts DJ, Seiden MV, Scadden DT, Rueda BR (2009) and others. CD133 expression defines a tumor initiating cell population in primary human ovarian cancer. Stem Cells 27(12):2875–2883
Deng S, Yang X, Lassus H, Liang S, Kaur S, Ye Q, Li C, Wang LP, Roby KF, Orsulic S et al (2010) Distinct expression levels and patterns of stem cell marker, aldehyde dehydrogenase isoform 1 (ALDH1), in human epithelial cancers. PLoS One 5(4):e10277
Dubeau L (2008) The cell of origin of ovarian epithelial tumours. Lancet Oncol 9(12):1191–1197
Emmanuel C, Gava N, Kennedy C, Balleine RL, Sharma R, Wain G, Brand A, Hogg R, Etemadmoghadam D, George J et al (2011) Comparison of expression profiles in ovarian epithelium in vivo and ovarian cancer identifies novel candidate genes involved in disease pathogenesis. PLoS One 6(3):e17617
Fargeas CA, Corbeil D, Huttner WB (2003) AC133 antigen, CD133, prominin-1, prominin-2, etc.: prominin family gene products in need of a rational nomenclature. Stem Cells 21(4):506–508
Farghaly S (2014) Cancer stem cells (Csc's): emerging concept and strategy for targeting progression of epithelial ovarian cancer. Enliven: Challenges Cancer Detect Ther 1(1):e001
Ferrandina G, Bonanno G, Pierelli L, Perillo A, Procoli A, Mariotti A, Corallo M, Martinelli E, Rutella S, Paglia A et al (2008) Expression of CD133-1 and CD133-2 in ovarian cancer. Int J Gynecol Cancer 18(3):506–514
Ferrandina G, Martinelli E, Petrillo M, Prisco MG, Zannoni G, Sioletic S, Scambia G (2009) CD133 antigen expression in ovarian cancer. BMC Cancer 9:221
Flesken-Nikitin A, Hwang CI, Cheng CY, Michurina TV, Enikolopov G, Nikitin AY (2013) Ovarian surface epithelium at the junction area contains a cancer-prone stem cell niche. Nature 495(7440):241–245
Foster R, Buckanovich RJ, Rueda BR (2013) Ovarian cancer stem cells: working towards the root of stemness. Cancer Lett 338(1):147–157
Friel AM, Sergent PA, Patnaude C, Szotek PP, Oliva E, Scadden DT, Seiden MV, Foster R, Rueda BR (2008) Functional analyses of the cancer stem cell-like properties of human endometrial tumor initiating cells. Cell Cycle 7(2):242–249
Gao MQ, Choi YP, Kang S, Youn JH, Cho NH (2010) CD24+ cells from hierarchically organized ovarian cancer are enriched in cancer stem cells. Oncogene 29(18):2672–2680
Gimenez-Bonafe P, Tortosa A, Perez-Tomas R (2009) Overcoming drug resistance by enhancing apoptosis of tumor cells. Curr Cancer Drug Targets 9(3):320–340

Guddati AK (2012) Ovarian cancer stem cells: elusive targets for chemotherapy. Med Oncol 29(5):3400–3408
Hanahan D, Weinberg RA (2011) Hallmarks of cancer: the next generation. Cell 144(5):646–674
Hope KJ, Jin L, Dick JE (2003) Human acute myeloid leukemia stem cells. Arch Med Res 34(6):507–514
Imrich S, Hachmeister M, Gires O (2012) EpCAM and its potential role in tumor-initiating cells. Cell Adh Migr 6(1):30–38
Jain AK, Allton K, Iacovino M, Mahen E, Milczarek RJ, Zwaka TP, Kyba M, Barton MC (2012) p53 regulates cell cycle and microRNAs to promote differentiation of human embryonic stem cells. PLoS Biol 10(2):e1001268
Jemal A, Siegel R, Ward E, Hao Y, Xu J, Thun MJ (2009) Cancer statistics, 2009. CA Cancer J Clin 59(4):225–249
Jemal A, Siegel R, Xu J, Ward E (2010) Cancer statistics, 2010. CA Cancer J Clin 60(5):277–300
Jordan CT (2004) Cancer stem cell biology: from leukemia to solid tumors. Curr Opin Cell Biol 16(6):708–712
Karst AM, Drapkin R (2010) Ovarian cancer pathogenesis: a model in evolution. J Oncol 2010:932371
Katz E, Skorecki K, Tzukerman M (2009) Niche-dependent tumorigenic capacity of malignant ovarian ascites-derived cancer cell subpopulations. Clin Cancer Res 15(1):70–80
Kim TH, Suh DH, Kim MK, Song YS (2014) Metformin against cancer stem cells through the modulation of energy metabolism: special considerations on ovarian cancer. Biomed Res Int 2014:132702
Kristiansen G, Sammar M, Altevogt P (2004) Tumour biological aspects of CD24, a mucin-like adhesion molecule. J Mol Histol 35(3):255–262
Kryczek I, Liu S, Roh M, Vatan L, Szeliga W, Wei S, Banerjee M, Mao Y, Kotarski J, Wicha MS et al (2012) Expression of aldehyde dehydrogenase and CD133 defines ovarian cancer stem cells. Int J Cancer 130(1):29–39
Kurman RJ, Shih IM (2010) The origin and pathogenesis of epithelial ovarian cancer: a proposed unifying theory. Am J Surg Pathol 34(3):433–443
Kusumbe AP, Bapat SA (2008) Ovarian stem cell biology and the emergence of ovarian cancer stem cells. In: Bapat S, Hoboken NJ (eds) Cancer stem cells. Wiley, Hoboken, pp 95–110
Kusumbe AP, Mali AM, Bapat SA (2009) CD133-expressing stem cells associated with ovarian metastases establish an endothelial hierarchy and contribute to tumor vasculature. Stem Cells 27(3):498–508
Landen CN Jr, Goodman B, Katre AA, Steg AD, Nick AM, Stone RL, Miller LD, Mejia PV, Jennings NB, Gershenson DM et al (2010) Targeting aldehyde dehydrogenase cancer stem cells in ovarian cancer. Mol Cancer Ther 9(12):3186–3199
Lawrenson K, Gayther SA (2009) Ovarian cancer: a clinical challenge that needs some basic answers. PLoS Med 6(2):e25
Ledermann JA, Raja FA, Fotopoulou C, Gonzalez-Martin A, Colombo N, Sessa C (2013) Newly diagnosed and relapsed epithelial ovarian carcinoma: ESMO clinical practice guidelines for diagnosis, treatment and follow-up. Ann Oncol 24(Suppl 6):vi24–vi32
Leeper K, Garcia R, Swisher E, Goff B, Greer B, Paley P (2002) Pathologic findings in prophylactic oophorectomy specimens in high-risk women. Gynecol Oncol 87(1):52–56
Leitao MM Jr, Chi DS (2009) Surgical management of recurrent ovarian cancer. Semin Oncol 36(2):106–111
Lim YC, Kang HJ, Kim YS, Choi EC (2012) All-trans-retinoic acid inhibits growth of head and neck cancer stem cells by suppression of Wnt/beta-catenin pathway. Eur J Cancer 48(17):3310–3318
Liu G, Yuan X, Zeng Z, Tunici P, Ng H, Abdulkadir IR, Lu L, Irvin D, Black KL, Yu JS (2006) Analysis of gene expression and chemoresistance of CD133+ cancer stem cells in glioblastoma. Mol Cancer 5:67
Luo L, Zeng J, Liang B, Zhao Z, Sun L, Cao D, Yang J, Shen K (2011) Ovarian cancer cells with the CD117 phenotype are highly tumorigenic and are related to chemotherapy outcome. Exp Mol Pathol 91(2):596–602

Marchitti SA, Brocker C, Stagos D, Vasiliou V (2008) Non-P450 aldehyde oxidizing enzymes: the aldehyde dehydrogenase superfamily. Expert Opin Drug Metab Toxicol 4(6):697–720
Miettinen M, Lasota J (2005) KIT (CD117): a review on expression in normal and neoplastic tissues, and mutations and their clinicopathologic correlation. Appl Immunohistochem Mol Morphol 13(3):205–220
Moreb JS (2008) Aldehyde dehydrogenase as a marker for stem cells. Curr Stem Cell Res Ther 3(4):237–246
Moulla A, Miliaras D, Sioga A, Kaidoglou A, Economou L (2013) The immunohistochemical expression of CD24 and CD171 adhesion molecules in borderline ovarian tumors. Pol J Pathol 64(3):180–184
Murdoch WJ, McDonnel AC (2002) Roles of the ovarian surface epithelium in ovulation and carcinogenesis. Reproduction 123(6):743–750
Ness RB, Cottreau C (1999) Possible role of ovarian epithelial inflammation in ovarian cancer. J Natl Cancer Inst 91(17):1459–1467
Orian-Rousseau V (2010) CD44, a therapeutic target for metastasising tumours. Eur J Cancer 46(7):1271–1277
Palmirotta R, Silvestris E, D'Oronzo S, Cardascia A, Silvestris F (2017) Ovarian cancer: novel molecular aspects for clinical assessment. Crit Rev Oncol Hematol 117:12–29
Pauli C, Munz M, Kieu C, Mack B, Breinl P, Wollenberg B, Lang S, Zeidler R, Gires O (2003) Tumor-specific glycosylation of the carcinoma-associated epithelial cell adhesion molecule EpCAM in head and neck carcinomas. Cancer Lett 193(1):25–32
Peng S, Maihle NJ, Huang Y (2010) Pluripotency factors Lin28 and Oct4 identify a sub-population of stem cell-like cells in ovarian cancer. Oncogene 29(14):2153–2159
Permuth-Wey J, Sellers TA (2009) Epidemiology of ovarian cancer. Methods Mol Biol 472:413–437
Pignata S, Scambia G, Ferrandina G, Savarese A, Sorio R, Breda E, Gebbia V, Musso P, Frigerio L, Del Medico P et al (2011) Carboplatin plus paclitaxel versus carboplatin plus pegylated liposomal doxorubicin as first-line treatment for patients with ovarian cancer: the MITO-2 randomized phase III trial. J Clin Oncol 29(27):3628–3635
Rehman K, Akash MSH (2017) Recent advances in lung regeneration. In: Pham PV (ed) Liver, lung and heart regeneration. Springer, Cham, pp 119–134
Rehman K, Iqbal MJ, Zahra N, Akash MS (2014) Liver stem cells: from preface to advancements. Curr Stem Cell Res Ther 9(1):10–21
Reya T, Morrison SJ, Clarke MF, Weissman IL (2001) Stem cells, cancer, and cancer stem cells. Nature 414(6859):105–111
Ruiz-Vela A, Aguilar-Gallardo C, Martinez-Arroyo AM, Soriano-Navarro M, Ruiz V, Simon C (2011) Specific unsaturated fatty acids enforce the transdifferentiation of human cancer cells toward adipocyte-like cells. Stem Cell Rev 7(4):898–909
Scott AM, Wolchok JD, Old LJ (2012) Antibody therapy of cancer. Nat Rev Cancer 12(4):278–287
Sebastian M, Kuemmel A, Schmidt M, Schmittel A (2009) Catumaxomab: a bispecific trifunctional antibody. Drugs Today (Barc) 45(8):589–597
Seimetz D, Lindhofer H, Bokemeyer C (2010) Development and approval of the trifunctional antibody catumaxomab (anti-EpCAM x anti-CD3) as a targeted cancer immunotherapy. Cancer Treat Rev 36(6):458–467
Sell S (2004) Stem cell origin of cancer and differentiation therapy. Crit Rev Oncol Hematol 51(1):1–28
Shank JJ, Yang K, Ghannam J, Cabrera L, Johnston CJ, Reynolds RK, Buckanovich RJ (2012) Metformin targets ovarian cancer stem cells in vitro and in vivo. Gynecol Oncol 127(2):390–397
Shi J, Zhou Z, Di W, Li N (2013) Correlation of CD44v6 expression with ovarian cancer progression and recurrence. BMC Cancer 13:182
Siegel R, Ward E, Brawley O, Jemal A (2011) Cancer statistics, 2011: the impact of eliminating socioeconomic and racial disparities on premature cancer deaths. CA Cancer J Clin 61(4):212–236
Silva IA, Bai S, McLean K, Yang K, Griffith K, Thomas D, Ginestier C, Johnston C, Kueck A, Reynolds RK et al (2011) Aldehyde dehydrogenase in combination with CD133 defines angiogenic ovarian cancer stem cells that portend poor patient survival. Cancer Res 71(11):3991–4001

Singh SK, Hawkins C, Clarke ID, Squire JA, Bayani J, Hide T, Henkelman RM, Cusimano MD, Dirks PB (2004) Identification of human brain tumour initiating cells. Nature 432(7015):396–401
Society AC (2014) Cancer facts and figures. Atlanta, GA, The Society
Soignet SL, Benedetti F, Fleischauer A, Parker BA, Truglia JA, Ra Crisp M, Warrell RP Jr (1998) Clinical study of 9-cis retinoic acid (LGD1057) in acute promyelocytic leukemia. Leukemia 12(10):1518–1521
Steffensen KD, Alvero AB, Yang Y, Waldstrom M, Hui P, Holmberg JC, Silasi DA, Jakobsen A, Rutherford T, Mor G (2011) Prevalence of epithelial ovarian cancer stem cells correlates with recurrence in early-stage ovarian cancer. J Oncol 2011:620523
Thiery JP, Acloque H, Huang RY, Nieto MA (2009) Epithelial-mesenchymal transitions in development and disease. Cell 139(5):871–890
Tomao F, Papa A, Rossi L, Strudel M, Vici P, Lo Russo G, Tomao S (2013) Emerging role of cancer stem cells in the biology and treatment of ovarian cancer: basic knowledge and therapeutic possibilities for an innovative approach. J Exp Clin Cancer Res 32:48
Tothill IE (2009) Biosensors for cancer markers diagnosis. Semin Cell Dev Biol 20(1):55–62
Whitworth JM, Londono-Joshi AI, Sellers JC, Oliver PJ, Muccio DD, Atigadda VR, Straughn JM Jr, Buchsbaum DJ (2012) The impact of novel retinoids in combination with platinum chemotherapy on ovarian cancer stem cells. Gynecol Oncol 125(1):226–230
Wikborn C, Pettersson F, Moberg PJ (1996) Delay in diagnosis of epithelial ovarian cancer. Int J Gynaecol Obstet 52(3):263–267
Wu Q, Guo R, Lin M, Zhou B, Wang Y (2011) MicroRNA-200a inhibits CD133/1+ ovarian cancer stem cells migration and invasion by targeting E-cadherin repressor ZEB2. Gynecol Oncol 122(1):149–154
Xu CX, Xu M, Tan L, Yang H, Permuth-Wey J, Kruk PA, Wenham RM, Nicosia SV, Lancaster JM, Sellers TA et al (2012) MicroRNA miR-214 regulates ovarian cancer cell stemness by targeting p53/Nanog. J Biol Chem 287(42):34970–34978
Yang H, Kong W, He L, Zhao JJ, O'Donnell JD, Wang J, Wenham RM, Coppola D, Kruk PA, Nicosia SV et al (2008) MicroRNA expression profiling in human ovarian cancer: miR-214 induces cell survival and cisplatin resistance by targeting PTEN. Cancer Res 68(2):425–433
Yin G, Alvero AB, Craveiro V, Holmberg JC, Fu HH, Montagna MK, Yang Y, Chefetz-Menaker I, Nuti S, Rossi M et al (2013) Constitutive proteasomal degradation of TWIST-1 in epithelial-ovarian cancer stem cells impacts differentiation and metastatic potential. Oncogene 32(1):39–49
Yu Z, Li Y, Fan H, Liu Z, Pestell RG (2012) miRNAs regulate stem cell self-renewal and differentiation. Front Genet 3:191–195
Zakaria N, Satar NA, Abu Halim NH, Ngalim SH, Yusoff NM, Lin J, Yahaya BH (2017) Targeting lung cancer stem cells: research and clinical impacts. Front Oncol 7:80
Zhang S, Balch C, Chan MW, Lai HC, Matei D, Schilder JM, Yan PS, Huang TH, Nephew KP (2008) Identification and characterization of ovarian cancer-initiating cells from primary human tumors. Cancer Res 68(11):4311–4320

Part II
Stem Cells for Genetic Diseases

Chapter 7
Mesenchymal Stem Cell Based Approaches for Systemic Lupus Erythematosus Immunotherapy

Xuexiao Jin, Shen Lu, and Yang Yao

Abbreviations

APCs	Antigen-presenting cells
B7-H1	B7 homolog 1
BAFF	B cell activating factor
BM	Bone marrow
Bregs	Regulatory B cells
DCs	Dendritic cells
HGF	Hepatocyte growth factor
HSCs	Hematopoietic stem cells
IDO	Indoleamine-2-3-dioxygenase
IFN-γ	Interferon gamma
MMF	Mycophenolate mofetil
MSCs	Mesenchymal stem cells
NK cells	Natural killer cells
NO	Nitric oxide
NSAIDs	Non-steroidal anti-inflammatory drugs
PGE2	Prostaglandin E2
SLE	Systemic lupus erythematosus

X. Jin (✉)
Institute of Immunology, Zhejiang University School of Medicine, Hangzhou, Zhejiang, China

Zhejiang University–University of Edinburgh Institute, Zhejiang University School of Medicine, Hangzhou, Zhejiang, China
e-mail: shirleyj@zju.edu.cn

S. Lu
Veritas Genetics Asia Inc., Hangzhou, Zhejiang, China

Y. Yao
Institute of Immunology, Zhejiang University School of Medicine, Hangzhou, Zhejiang, China
e-mail: 0916036@zju.edu.cn

P. V. Pham, A. El-Hashash (eds.), *Stem Cells for Cancer and Genetic Disease Treatment*, Stem Cells in Clinical Applications, https://doi.org/10.1007/978-3-319-98065-2_7

SLEDAI	Systemic lupus erythematosus disease activity index
TGF-β	Transforming growth factor-β
TH cells	T helper cells
TNF-α	Tumor necrosis factor alpha
Tregs	Regulatory T cells
UC	Umbilical cord

7.1 Brief Introduction of SLE

Systemic lupus erythematosus (SLE) is a complex and heterogeneous autoimmune disease characterized by a relapsing course with periods of flares, alternating periods of remission and by highly heterogeneous clinical manifestations with multi-organ involvement (Rahman and Isenberg 2008). The pathogenesis of SLE is complicated, which relies on a complex interplay of genetic, endogenous, and environmental factors, resulting in a breakdown of self-tolerance, production of autoantibody, deposition of immune complex, and organ damage (Arbuckle et al. 2003).

SLE predominantly affects women of 20–40 years age, with a female: male ratio of 9:1 (Danchenko et al. 2006). The clinical symptoms of SLE are diverse, including fatigue, oral ulcerations, arthralgias, and even mortal hematological, renal and neurological disease. Disease activity fluctuates with periods of flares and remission.

Currently, treatments for SLE mainly include NSAIDs, antimalarial drugs, and glucocorticoids. For patients with severe symptoms such as organ dysfunction, immunosuppressive agents such as methotrexate, azathioprine, mycophenolate mofetil (MMF), cyclophosphamide, and cyclosporine are additionally used (Xiong and Lahita 2014). But the limitation of these mainstay treatments in refractory SLE is unneglectable. In last decades, progresses in understanding the immune-pathogenesis of SLE and the development of immunotherapies for the disease have been achieved. The new strategies of targeted immunotherapy have shown promising result in SLE treatment.

7.2 Immunotherapies for SLE

Although treatments with current pharmaceutical products can control the symptoms and prevent organ damage, the benefits on severe cases are limited. Also, they often result in serious side effects, including infection, secondary malignancy, and relapses of the disease after drug withdrawal (Bernatsky et al. 2006). Development of new therapeutic strategies with more beneficial effect and less toxicity is under urge.

It is known that abnormalities in lymphocytes distribution and accompanying dysregulation of inflammatory cytokines are crucial events in pathogenesis of SLE (Gatto et al. 2013). Most patients with autoimmune disorders show decreased number or function of regulatory cells. Altering the distribution of lymphocytes by immunotherapeutic strategy may be a potential treatment in autoimmune diseases. For example, transplantation of mesenchymal stem cells (MSCs), which have a strong regulating function, is beneficial to a specific subpopulation of lupus patients. Regulatory T cells (Tregs) are also a type of regulator cell with high efficiency to prevent the excessive immune activation, while depleting B cell with monoclonal antibody such as rituximab is a very good alternative and safe therapy in SLE as well (Liao et al. 2015).

Among the immunotherapeutic strategies, the treatment with MSCs has been tested in many autoimmune disorders such as refractory SLE, Type I diabetes, systemic multiple sclerosis, Crohn's disease, and arthritis. The immune-regulatory and tissue regeneration properties of MSCs as well as the low immunogenic potential make MSCs a very promising new immunotherapy for severe refractory SLE.

7.3 Abnormal MSCs in SLE Progression

Mesenchymal stem cells (MSCs) are multi-potent, non-hematopoietic progenitor cells that can be isolated from multiple sources including placenta, umbilical cord (UC), adipose tissue, teeth, and menstrual fluid (Hass et al. 2011).

MSCs have the ability to self-renew and give rise to different cell lineages (Pal et al. 2009; Shi et al. 2010) and also can enhance hematopoiesis, possess immune-modulatory properties, and inhibit T cell proliferation in vitro (Gu et al. 2012b; Schurgers et al. 2010).

SLE is a chronic autoimmune disease characterized by multi-organ involvement and remarkable variability in clinical presentation (Grant 2001). Dysregulated activation of both B and T cells and the aberrant production of pro-inflammatory cytokines have been critically involved in the initiation and progress of tissue pathology and organ damage in SLE (Ren et al. 2008). Many studies suggest that MSCs, a key component of hematopoietic microenvironment are abnormal during SLE progression (Sun et al. 2009).

The studies of Sun et al. concentrate on the phenotype, karyotype, cytokine expression, and hematopoietic support of MSCs. They found that MSCs from both SLE patients and normal people have the ability to be cultured in vitro but MSCs from SLE patients show impaired capability of proliferation and generation compared to normal MSCs (Sun et al. 2007; Tang et al. 2013).

A study found that BMP/TGF-β and MAPK signaling pathways are differentially expressed in the SLE BM-MSCs (Tang et al. 2012). It supports previous studies that abnormal cytokine secretion in patients with SLE, including IL-4, IL-6 and other cytokines, will lead to immune imbalance (Gu et al. 2012a). In addition, MSCs have been reported to suppress immune system by inhibiting the effector function of neutrophils and natural killer cells, meanwhile mediating their effects on Treg cells (Uccelli et al. 2006). As well as the direct effects of MSCs on immune cell functions, MSCs also have the capability of regulating the recruitment of circulating lymphocytes indirectly via interactions with vascular endothelium (Luu et al. 2013). MSCs can directly affect the fate and function of many distinct lymphocyte subsets primarily through the production of soluble anti-inflammatory cytokines.

Based on evidences above, we conclude that MSCs from SLE patients have a defective phenotype and functional defect in cytokine production which may lead to hyper-activation and dysfunction of immune cells. Whether the defective is primary or secondary is still controversial.

7.4 MSCs-Based Therapy in SLE

7.4.1 MSCs-Based Therapy in Animal Models of SLE

Treatment with MSCs has been tested in a variety of autoimmune disease models, including murine SLE model. MRL/lpr mouse is a classic murine SLE model, of which Fas gene was mutated and the lupus-like symptoms appear spontaneously at 12-week-old age. Sun et al. hypothesized that impaired MSCs derived from BM (BM-MSCs) could accelerate the process of SLE in human and murine. Therefore, allogeneic mouse or human BM-MSCs or UC (UC-MSCs) were proved to be effective in reducing disease severity in the lpr mutant mouse model of SLE, assessed by proteinuria, serum auto-reactive antibody levels, the degree of renal pathology, or survival data (Zhou et al. 2008; Zhang et al. 2014). Moreover, human MSCs from healthy donors reduced the proliferation of T lymphocytes from MRL/lpr mice in a dose-dependent manner.

But the studies based on the NZB/WF1 strain, which is considered the closest murine model to human SLE, has shown diverging results. The study of Gu et al. suggested that MSCs are effective in treating established SLE in both MRL/lpr and (NZB/NZW) F1 mice (Gu et al. 2012b). They found that the early-stage treatment is more beneficial than the advanced-stage treatment, by measuring survival rates and proteinuria production. The result of Youd and Hong, on the contrary, demonstrated that intervention with allogeneic MSCs from BALB/c actually aggravated the disease development in (NZB × NZW) F1 mice, with more severe immune complex deposition in the kidney, increased proteinuria, and elevated autoantibody level (Youd et al. 2010; Choi et al. 2012).

7.4.2 MSCs-Based Therapy in Human SLE

A pilot study applied systemic allogeneic BM-MSCs administration to four patients with active renal disease that were unresponsive to standard immunosuppressive therapy (Sun et al. 2009). It was encouraging that improved renal function and reduced SLE disease activity were found in treated patients, such that patients were in remission for up to 12 months after treatment with no side effect (Sun et al. 2009; Li et al. 2010; Stock et al. 2010).

Based on this result, 15 patients with refractory SLE disease received infusion of allogeneic BM-MSCs from non-HLA matched healthy family members in a Phase I trial. Clinical and serological changes were quite dramatic, that Systemic Lupus Erythematosus Disease Activity Index (SLEDAI), serum albumin, proteinuria, serum complement, and anti-dsDNA antibodies improved significantly (Wang et al. 2014a).

Alchi et al. reported that the 5-year-overall-survival rate in their 28 SLE patients was 81% (5 patients died in 2 years after treatment, 3 deaths were caused by infections, 1 progressive SLE, and 1 caused by autoimmune hemolytic anemia). Burt et al. observed a 50% disease-free survival rate in 50 SLE patients at 5 years (2 patients died after mobilization: one active lupus and one from mucormycosis) (Burt et al. 2006).

These data proved that MSCs treatment has efficient clinical capability in severe SLE with no significant toxicity which strongly suggests that MSCs treatment may have substantial benefit for patients suffering from SLE. As a result, several large-scale, prospective, double-blind, multicenter trials are now under way to optimize the treatment protocol.

7.5 Mechanisms

Although the in vitro and in vivo data of the therapeutic effects of MSCs treatment in SLE appears promising, the underlying protective mechanisms still remain elusive. Current understanding of the mechanisms is summarized below (Fig. 7.1).

7.5.1 MSCs Homing and Cytokine Production

After a single intravenous (i.v.) infusion of autologous or allogeneic MSCs in baboons, levels of tissue engraftment ranges from 0.1 to 2.7% (Devine et al. 2003). In NZB/W lupus mice treated with 1×10^6 human UC-MSCs i.v., MSCs were detected in kidneys 2 weeks after infusion (Chang et al. 2011). This data indicate that MSCs homing happens after infusion while the homing efficiency is limited.

Previous studies have shown that MSCs can alter the functions of immune cells by cell-to-cell contact and secretion of soluble cytokines such as IL-10, transforming growth factor-β (TGF-β), prostaglandin E2 (PGE2), nitric oxide (NO), and indoleamine-2-3-dioxygenase (IDO) (Fan et al. 2016). These cytokines mediate the interaction between MSCs and other immune cells. And IFN- γ, alone or together with TNF-α, IL-1α, or IL-1β, is required for MSCs differentiation and production of high levels of these cytokines (Ren et al. 2008; Figueroa et al. 2014). A recent study found that the serum IFN-γ level of SLE patients can predict the outcome of MSCs infusion treatment (Wang et al. 2017).

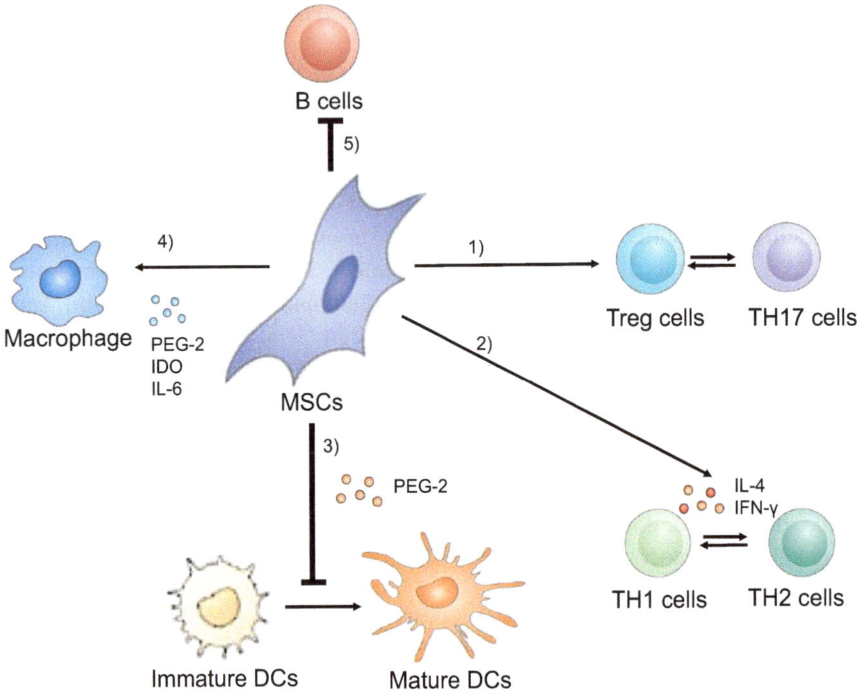

Fig. 7.1 Immunomodulatory role of MSCs on immune cells in SLE. (1) MSCs have a role of promoting generation of functional Tregs and reducing TH17 cells significantly. (2) MSCs promote the conversion from a T helper (TH)2 humoral response to a TH1 cellular immune response through the modulation of IL-4 and IFN-γ levels in TH cells. (3) MSCs inhibit dendritic cells maturation by secretion of PGE2. (4) MSCs also promote polarization of monocytes towards an M2 phenotype. (5) MSCs inhibit B cell proliferation and plasma cell differentiation. Abbreviations: *DCs* dendritic cells, *PGE2* prostaglandin E2, *MSCs* mesenchymal stem cells, *IDO* indoleamine 2,3,dioxygenase, *TH cells* T helper cells, *IFN* interferon

7.5.2 *MSCs Regulate Distribution and Function of T Cell Subsets*

7.5.2.1 Balancing of TH1 and TH2 Cells

There are studies suggesting that MSCs therapy promotes the conversion from humoral response of T helper (TH) 2 to cellular immune response of TH1 through the modulation of IL-4 and IFN-γ levels in TH cells, which might ameliorate SLE. The intraperitoneal infusion of human BM-MSCs in MRL/lpr mice affects function of peripheral T cells by decreasing production of IL-4 and increasing IFN-γ (Zhou et al. 2008). Similar findings have been reported that transplantation of UC-MSCs in refractory human SLE promotes polarization of periperal T cells to the TH1 phenotype (Sun et al. 2010). However, Aggarwal et al. showed opposite effects using human MSCs (Aggarwal and

Pittenger 2005), and Chang et al. found that UC-MSCs transplantation in NZB/W mice associated with an increase in TH2 cells and a disease improvement (Chang et al. 2011). Those ambivalent results suggest the complexity of both the effects of MSCs and the immune-pathogenesis of SLE.

7.5.2.2 Balancing of Treg and TH17 Cells

Studies proved there is a deficit of regulatory T cells (Tregs) in human and murine SLE, while an increased activity of TH17 pro-inflammatory cells (Tao et al. 2017). It is known that MSCs have a role of promoting generation of functional Tregs both in vitro and in vivo (Gonzalez et al. 2009; Crispin and Tsokos 2010). In MLR/lpr mice, the transplantation of MSCs from different sources (BM, UC, or exfoliated deciduous teeth) restored Treg cells activity and significantly reduced TH17 cells (Sun et al. 2009). Transplantation of either allogeneic or autologous MSCs derived from BM or UC increased Treg cell numbers in human SLE as well, which suggest this might be one of the mechanisms of MSCs-mediated improvement of SLE (Sun et al. 2010).

Besides, in vitro and in vivo evidences suggest that BM-MSCs can also inhibit T follicular cells differentiation and cytokine production (Yang et al. 2017).

7.5.3 Effects on Dendritic/Macrophage Cells

7.5.3.1 MSCs Modulate DCs at Multiple Levels

Dendritic cells (DCs) are the most potent antigen-presenting cells (APCs). Dendritic cells can activate both naïve and memory periphery T cells and help maintaining the balance between immune response and tolerance. There are different types of DCs subsets contributing in the balancing of immune system. As DCs secrete large amounts of IFN-α, they play an important role in the pathogenesis of SLE. And the SLE patient serum with enriched IFN-α can differentiate monocytes into DCs (Blanco et al. 2001). Moreover, deficiency of IFN-α receptor in mouse leads to reduced symptoms of SLE (Santiago-Raber et al. 2003).

As reviewed by Spaggiari et al., most studies that address the interaction between MSCs and APCs have demonstrated that MSCs modulate dendritic cells (DCs) at multiple levels. MSCs alter the phenotype, cytokine release, differentiation and maturation of DCs, and compromise their antigen presentation ability (Spaggiari and Moretta 2013). Also, a study investigated the effect of MSCs on different monocyte subpopulations (classical, intermediate, and non-classical) (Laranjeira et al. 2015; Najar et al. 2016). MSCs differentially modulate the cytokine secretion profile of these monocytes subsets. As a result, DCs are unable to efficiently prime T cells for a robust immune response. PGE2 is a potent inducer of IL-10, and both factors have a pivotal role in the cross-regulation of DCs. Thus, PGE2-induced IL-10 is a key regulator of the bone marrow DC (BM-DC) pro-inflammatory response (Harizi and Gualde 2006).

7.5.3.2 MSCs Alter M1/M2 Balancing of Macrophages

Macrophages can be derived from circulating inflammatory or resident monocytes, which are recruited to tissue sites of infection and inflammation. They can be high plasticity in response to environmental signals (Murray et al. 2014). Thus, different populations of macrophages with distinct functions may arise, including the pro-inflammatory subtype (M1, classically activated macrophages) and the anti-inflammatory subtype (M2, alternatively activated macrophages) (Martinez et al. 2008). IDO activity in MSCs is involved in the generation of anti-inflammatory M2 macrophages. These M2 cells are involved in T cell inhibition in an IL-10-independent manner, thereby amplifying the immunosuppressive effects of MSCs (Francois et al. 2012). Interestingly, through the constitutive secretion of IL-6, MSCs induce M2 macrophages that are characterized by concomitant enhanced production of IL-10 (Melief et al. 2013). The roles of IL-6 and PGE2 on monocytes are not considered contradictory but work in a synergistic manner. IL-6 is involved in M2 macrophage generation, whereas PGE2 selectively interferes with the generation of immature DCs. Conversely, IL-6-dependent PGE2 secretion is correlated with the immunosuppressive activity of MSCs (Bouffi et al. 2010). Indeed, PGE2 regulates the production of a wide array of cytokines. PGE2 upregulates both IL-10 and IL-6 in activated murine macrophages (Williams et al. 2000).

7.5.4 Effects on B Cells

7.5.4.1 Dysfunction of B Cells in SLE

Hyper-reactivity of polyclonal auto-reactive B cells is the main pathogenic characteristic of SLE. These auto-reactive B cells secrete pathogenic antibodies and produce cytokines which lead to the activation of T cells (Marino and Grey 2012). In murine and human SLE, these cells show a loss of B cell tolerance (Renaudineau et al. 2004). Excessive autoantibody secretion leads to the formation and deposition of immune complex in internal organs, which ultimately causes serious inflammation and irreversible organ damage (Mandik-Nayak et al. 2008).

In lupus mice and SLE patients, dysfunction of B cell tolerance, specifically related to negative selection, and lowered activation threshold have been characterized. This may explain the persistence of auto-reactive B cells and fluctuation of disease activity. Studies show that defects in Fcγ receptors, lyn or BCR/TLR may promote auto-reactive B cell activation (Huck et al. 2001). Besides, anti-double stranded DNA antibodies produced by auto-reactive B cells have been proven to facilitate lupus nephritis pathogenesis (Bernard et al. 2014). In recent years, monoclonal antibodies targeting B cell have shown delightful therapeutic effects in SLE patients, and further clinical studies are under way. Dysfunction of B cell becomes ameliorated after MSCs application in patients.

7.5.4.2 Interaction Between MSCs and B Cells

As is known, MSCs can regulate the functions of immune cells by both cell-to-cell contact and secretion of soluble cytokines, including TGF-β, IL-10, PGE2, NO, and IDO. For B cells, interestingly, current studies suggest that the interaction of B cells and MSCs mostly depends on the culture microenvironment. In some circumstances, B cells may alter characteristics and behaviors of MSCs by directly interacting with them. Other immune cells, including T cells and DCs, intermediate between the MSCs and B cells and indirectly regulate the functions of these cells (Fan et al. 2016).

The interplay between MSCs and B cells is complicated. In vitro environments have shown that MSCs can specifically inhibit B cell proliferation, plasma cell differentiation and promote Breg cell proliferation while cultured with multiple types of immune cells (Rosado et al. 2015). Some specific cytokines, including IFN-γ and BAFF, are involved. According to the data of Rosado, M.M. et al., MSCs can suppress the IFN-γ secretion capacity of T cells in this process. And IFN-γ is a type of regulator of B cell proliferation and differentiation; the IFN-γ secretion of T cells partially relies on cytokine signals of activated B cells. Thus, MSCs inhibit B cell proliferation and plasma cell differentiation indirectly by inhibiting IFN-γ secretion of T cells. Similarly, MSCs can inhibit B cell function via downregulation of BAFF secretion by DCs.

Also, evidences show that cell-to-cell contact also plays a role in MSCs–B cell interaction. Cell-to-cell contact between MSCs and B cells is required for B7-H1 and Gal-9 to deliver inhibitory signal to B cells (Wang et al. 2014b).

7.6 Issues in MSCs Treatment of SLE

Preclinical and clinical trial data have shown efficacy in MSCs treatment of SLE. But still, a lot of issues are limiting the use of MSCs as a therapeutic intervention for SLE.

First of all, MSCs are a heterogeneous population of stromal cells which currently have no specific markers. The limited knowledge about MSCs hampers the standardization of MSCs culture protocol, which minimizes changes in MSCs phenotype. A lack of standardization in MSCs isolation and amplification protocol and understanding of basic MSCs biology is limiting the development of this method, which results in the conflicting outcomes. So, to fully understand the heterogeneity and efficacy of different MSCs populations appears to be one of the most important subjects.

Secondly, studies have shown that MSCs have a very low homing efficiency after systemically infused. Most of them have been trapped in the lungs and the liver (Munir and McGettrick 2015). To maximize the treatment efficacy, standard delivery method of MSCs therapy is required.

Furthermore, the mechanism of MSCs interacting with immune cells is not fully revealed, for example, neither the exact function of MSCs in TH cells balancing nor the molecular events of the cell-to-cell contact between MSCs and B cells are fully understood.

Lastly, the long-term effects of therapy are not clearly understood. As the speed by which MSCs are cleared from the patient is unknown, the study of long-term side effects of treatment is even more elusive. Also, the duration of the beneficial effects of this treatment is currently unknown. If repeat interventions are required to maintain effects, and whether the condition of the patient get worsen again after a few years are remain to be studied.

7.7 Conclusion

In the past decades, fascinating in vitro and in vivo data have shown the immunosuppressant properties of MSCs, and their potential to be used in clinical practice as a new therapeutic option of SLE treatment. In this chapter, we show a defect of MSCs in the SLE patients. The results of the clinical trials with MSCs therapy in severe SLE are undoubtedly encouraging. However, the mechanisms were not fully addressed so far. Further understanding of the interaction between MSCs and immune cells, or inflammatory tissue microenvironments and their underlying molecular events will be crucial for development of this potential therapeutic application of MSCs-based immunotherapy. Moreover, many issues, including the standardization of MSCs isolation and amplification, and cell doses, frequency, route of administration, etc., remain to be addressed. Furthermore, to claim the safety and long-term effect of this promising SLE immunotherapeutic strategy, a larger-scale and long-term clinical trial is required. We have seen much inspiring outcomes, but clearly there is still a long way to go.

References

Aggarwal S, Pittenger MF (2005) Human mesenchymal stem cells modulate allogeneic immune cell responses. Blood 105:1815–1822

Arbuckle MR, McClain MT, Rubertone MV, Scofield RH, Dennis GJ, James JA, Harley JB (2003) Development of autoantibodies before the clinical onset of systemic lupus erythematosus. N Engl J Med 349:1526–1533

Bernard M, Dieude M, Yang B, Hamelin K, Underwood K, Hebert MJ (2014) Autophagy fosters myofibroblast differentiation through MTORC2 activation and downstream upregulation of CTGF. Autophagy 10:2193–2207

Bernatsky S, Boivin JF, Joseph L, Manzi S, Ginzler E, Gladman DD, Urowitz M, Fortin PR, Petri M, Barr S, Gordon C, Bae SC, Isenberg D, Zoma A, Aranow C, Dooley MA, Nived O, Sturfelt G, Steinsson K, Alarcon G, Senecal JL, Zummer M, Hanly J, Ensworth S, Pope J, Edworthy S, Rahman A, Sibley J, El-Gabalawy H, McCarthy T, St Pierre Y, Clarke A, Ramsey-Goldman R (2006) Mortality in systemic lupus erythematosus. Arthritis Rheum 54:2550–2557

Blanco P, Palucka AK, Gill M, Pascual V, Banchereau J (2001) Induction of dendritic cell differentiation by IFN-alpha in systemic lupus erythematosus. Science 294:1540–1543

Bouffi C, Bony C, Courties G, Jorgensen C, Noel D (2010) IL-6-dependent PGE2 secretion by mesenchymal stem cells inhibits local inflammation in experimental arthritis. PLoS One 5:e14247

Burt RK, Traynor A, Statkute L, Barr WG, Rosa R, Schroeder J, Verda L, Krosnjar N, Quigley K, Yaung K, Villa Bs M, Takahashi M, Jovanovic B, Oyama Y (2006) Nonmyeloablative hematopoietic stem cell transplantation for systemic lupus erythematosus. JAMA 295:527–535

Chang JW, Hung SP, Wu HH, Wu WM, Yang AH, Tsai HL, Yang LY, Lee OK (2011) Therapeutic effects of umbilical cord blood-derived mesenchymal stem cell transplantation in experimental lupus nephritis. Cell Transplant 20:245–257

Choi EW, Shin IS, Park SY, Park JH, Kim JS, Yoon EJ, Kang SK, Ra JC, Hong SH (2012) Reversal of serologic, immunologic, and histologic dysfunction in mice with systemic lupus erythematosus by long-term serial adipose tissue-derived mesenchymal stem cell transplantation. Arthritis Rheum 64:243–253

Crispin JC, Tsokos GC (2010) Interleukin-17-producing T cells in lupus. Curr Opin Rheumatol 22:499–503

Danchenko N, Satia JA, Anthony MS (2006) Epidemiology of systemic lupus erythematosus: a comparison of worldwide disease burden. Lupus 15:308–318

Devine SM, Cobbs C, Jennings M, Bartholomew A, Hoffman R (2003) Mesenchymal stem cells distribute to a wide range of tissues following systemic infusion into nonhuman primates. Blood 101:2999–3001

Fan L, Hu C, Chen J, Cen P, Wang J, Li L (2016) Interaction between mesenchymal stem cells and B-cells. Int J Mol Sci 17

Figueroa FE, Cuenca Moreno J, La Cava A (2014) Novel approaches to lupus drug discovery using stem cell therapy. Role of mesenchymal-stem-cell-secreted factors. Expert Opin Drug Discov 9:555–566

Francois M, Romieu-Mourez R, Li M, Galipeau J (2012) Human MSC suppression correlates with cytokine induction of indoleamine 2,3-dioxygenase and bystander M2 macrophage differentiation. Mol Ther 20:187–195

Gatto M, Zen M, Ghirardello A, Bettio S, Bassi N, Iaccarino L, Punzi L, Doria A (2013) Emerging and critical issues in the pathogenesis of lupus. Autoimmun Rev 12:523–536

Gonzalez MA, Gonzalez-Rey E, Rico L, Buscher D, Delgado M (2009) Treatment of experimental arthritis by inducing immune tolerance with human adipose-derived mesenchymal stem cells. Arthritis Rheum 60:1006–1019

Grant EC (2001) Systemic lupus erythematosus. Lancet 358:586

Gu Z, Cao X, Jiang J, Li L, Da Z, Liu H, Cheng C (2012a) Upregulation of p16INK4A promotes cellular senescence of bone marrow-derived mesenchymal stem cells from systemic lupus erythematosus patients. Cell Signal 24:2307–2314

Gu F, Molano I, Ruiz P, Sun L, Gilkeson GS (2012b) Differential effect of allogeneic versus syngeneic mesenchymal stem cell transplantation in MRL/lpr and (NZB/NZW)F1 mice. Clin Immunol 145:142–152

Harizi H, Gualde N (2006) Pivotal role of PGE2 and IL-10 in the cross-regulation of dendritic cell-derived inflammatory mediators. Cell Mol Immunol 3:271–277

Hass R, Kasper C, Bohm S, Jacobs R (2011) Different populations and sources of human mesenchymal stem cells (MSC): a comparison of adult and neonatal tissue-derived MSC. Cell Commun Signal 9:12

Huck S, Le Corre R, Youinou P, Zouali M (2001) Expression of B cell receptor-associated signaling molecules in human lupus. Autoimmunity 33:213–224

Laranjeira P, Gomes J, Pedreiro S, Pedrosa M, Martinho A, Antunes B, Ribeiro T, Santos F, Domingues R, Abecasis M, Trindade H, Paiva A (2015) Human bone marrow-derived mesenchymal stromal cells differentially inhibit cytokine production by peripheral blood monocytes subpopulations and myeloid dendritic cells. Stem Cells Int 2015:819084

Li K, Han Q, Yan X, Liao L, Zhao RC (2010) Not a process of simple vicariousness, the differentiation of human adipose-derived mesenchymal stem cells to renal tubular epithelial cells plays an important role in acute kidney injury repairing. Stem Cells Dev 19:1267–1275

Liao J, Chang C, Wu H, Lu Q (2015) Cell-based therapies for systemic lupus erythematosus. Autoimmun Rev 14:43–48

Luu NT, McGettrick HM, Buckley CD, Newsome PN, Rainger GE, Frampton J, Nash GB (2013) Crosstalk between mesenchymal stem cells and endothelial cells leads to downregulation of cytokine-induced leukocyte recruitment. Stem Cells 31:2690–2702

Mandik-Nayak L, Ridge N, Fields M, Park AY, Erikson J (2008) Role of B cells in systemic lupus erythematosus and rheumatoid arthritis. Curr Opin Immunol 20:639–645

Marino E, Grey ST (2012) B cells as effectors and regulators of autoimmunity. Autoimmunity 45:377–387

Martinez FO, Sica A, Mantovani A, Locati M (2008) Macrophage activation and polarization. Front Biosci 13:453–461

Melief SM, Geutskens SB, Fibbe WE, Roelofs H (2013) Multipotent stromal cells skew monocytes towards an anti-inflammatory function: the link with key immunoregulatory molecules. Haematologica 98:e121–e122

Munir H, McGettrick HM (2015) Mesenchymal stem cell therapy for autoimmune disease: risks and rewards. Stem Cells Dev 24:2091–2100

Murray PJ, Allen JE, Biswas SK, Fisher EA, Gilroy DW, Goerdt S, Gordon S, Hamilton JA, Ivashkiv LB, Lawrence T, Locati M, Mantovani A, Martinez FO, Mege JL, Mosser DM, Natoli G, Saeij JP, Schultze JL, Shirey KA, Sica A, Suttles J, Udalova I, van Ginderachter JA, Vogel SN, Wynn TA (2014) Macrophage activation and polarization: nomenclature and experimental guidelines. Immunity 41:14–20

Najar M, Raicevic G, Fayyad-Kazan H, Bron D, Toungouz M, Lagneaux L (2016) Mesenchymal stromal cells and immunomodulation: a gathering of regulatory immune cells. Cytotherapy 18:160–171

Pal R, Hanwate M, Jan M, Totey S (2009) Phenotypic and functional comparison of optimum culture conditions for upscaling of bone marrow-derived mesenchymal stem cells. J Tissue Eng Regen Med 3:163–174

Rahman A, Isenberg DA (2008) Systemic lupus erythematosus. N Engl J Med 358:929–939

Ren G, Zhang L, Zhao X, Xu G, Zhang Y, Roberts AI, Zhao RC, Shi Y (2008) Mesenchymal stem cell-mediated immunosuppression occurs via concerted action of chemokines and nitric oxide. Cell Stem Cell 2:141–150

Renaudineau Y, Pers JO, Bendaoud B, Jamin C, Youinou P (2004) Dysfunctional B cells in systemic lupus erythematosus. Autoimmun Rev 3:516–523

Rosado MM, Bernardo ME, Scarsella M, Conforti A, Giorda E, Biagini S, Cascioli S, Rossi F, Guzzo I, Vivarelli M, Dello Strologo L, Emma F, Locatelli F, Carsetti R (2015) Inhibition of B-cell proliferation and antibody production by mesenchymal stromal cells is mediated by T cells. Stem Cells Dev 24:93–103

Santiago-Raber ML, Baccala R, Haraldsson KM, Choubey D, Stewart TA, Kono DH, Theofilopoulos AN (2003) Type-I interferon receptor deficiency reduces lupus-like disease in NZB mice. J Exp Med 197:777–788

Schurgers E, Kelchtermans H, Mitera T, Geboes L, Matthys P (2010) Discrepancy between the in vitro and in vivo effects of murine mesenchymal stem cells on T-cell proliferation and collagen-induced arthritis. Arthritis Res Ther 12:R31

Shi Y, Hu G, Su J, Li W, Chen Q, Shou P, Xu C, Chen X, Huang Y, Zhu Z, Huang X, Han X, Xie N, Ren G (2010) Mesenchymal stem cells: a new strategy for immunosuppression and tissue repair. Cell Res 20:510–518

Spaggiari GM, Moretta L (2013) Interactions between mesenchymal stem cells and dendritic cells. Adv Biochem Eng Biotechnol 130:199–208

Stock P, Bruckner S, Ebensing S, Hempel M, Dollinger MM, Christ B (2010) The generation of hepatocytes from mesenchymal stem cells and engraftment into murine liver. Nat Protoc 5:617–627

Sun LY, Zhang HY, Feng XB, Hou YY, Lu LW, Fan LM (2007) Abnormality of bone marrow-derived mesenchymal stem cells in patients with systemic lupus erythematosus. Lupus 16:121–128

Sun L, Akiyama K, Zhang H, Yamaza T, Hou Y, Zhao S, Xu T, Le A, Shi S (2009) Mesenchymal stem cell transplantation reverses multiorgan dysfunction in systemic lupus erythematosus mice and humans. Stem Cells 27:1421–1432

Sun L, Wang D, Liang J, Zhang H, Feng X, Wang H, Hua B, Liu B, Ye S, Hu X, Xu W, Zeng X, Hou Y, Gilkeson GS, Silver RM, Lu L, Shi S (2010) Umbilical cord mesenchymal stem cell transplantation in severe and refractory systemic lupus erythematosus. Arthritis Rheum 62:2467–2475

Tang Y, Ma X, Zhang H, Gu Z, Hou Y, Gilkeson GS, Lu L, Zeng X, Sun L (2012) Gene expression profile reveals abnormalities of multiple signaling pathways in mesenchymal stem cell derived from patients with systemic lupus erythematosus. Clin Dev Immunol 2012:826182

Tang Y, Xie H, Chen J, Geng L, Chen H, Li X, Hou Y, Lu L, Shi S, Zeng X, Sun L (2013) Activated NF-kappaB in bone marrow mesenchymal stem cells from systemic lupus erythematosus patients inhibits osteogenic differentiation through downregulating Smad signaling. Stem Cells Dev 22:668–678

Tao JH, Cheng M, Tang JP, Liu Q, Pan F, Li XP (2017) Foxp3, regulatory T cell, and autoimmune diseases. Inflammation 40:328–339

Uccelli A, Moretta L, Pistoia V (2006) Immunoregulatory function of mesenchymal stem cells. Eur J Immunol 36:2566–2573

Wang D, Li J, Zhang Y, Zhang M, Chen J, Li X, Hu X, Jiang S, Shi S, Sun L (2014a) Umbilical cord mesenchymal stem cell transplantation in active and refractory systemic lupus erythematosus: a multicenter clinical study. Arthritis Res Ther 16:R79

Wang H, Qi F, Dai X, Tian W, Liu T, Han H, Zhang B, Li H, Zhang Z, Du C (2014b) Requirement of B7-H1 in mesenchymal stem cells for immune tolerance to cardiac allografts in combination therapy with rapamycin. Transpl Immunol 31:65–74

Wang D, Wang S, Huang S, Zhang Z, Yuan X, Feng X, Lu L, Sun L (2017) Serum IFN-gamma predicts the therapeutic effect of mesenchymal stem cells transplantation in systemic lupus erythematosus patients. Stem Cells Transl Med 6:1777–1785

Williams JA, Pontzer CH, Shacter E (2000) Regulation of macrophage interleukin-6 (IL-6) and IL-10 expression by prostaglandin E2: the role of p38 mitogen-activated protein kinase. J Interf Cytokine Res 20:291–298

Xiong W, Lahita RG (2014) Pragmatic approaches to therapy for systemic lupus erythematosus. Nat Rev Rheumatol 10:97–107

Yang X, Yang J, Li X, Ma W, Zou H (2018) Bone marrow-derived mesenchymal stem cells inhibit T follicular helper cell in lupus-prone mice. Lupus 27(1):49–59. https://doi.org/10.1177/0961203317711013

Youd M, Blickarz C, Woodworth L, Touzjian T, Edling A, Tedstone J, Ruzek M, Tubo R, Kaplan J, Lodie T (2010) Allogeneic mesenchymal stem cells do not protect NZBxNZW F1 mice from developing lupus disease. Clin Exp Immunol 161:176–186

Zhang Y, Xia Y, Ni S, Gu Z, Liu H (2014) Transplantation of umbilical cord mesenchymal stem cells alleviates pneumonitis of MRL/lpr mice. J Thorac Dis 6:109–117

Zhou K, Zhang H, Jin O, Feng X, Yao G, Hou Y, Sun L (2008) Transplantation of human bone marrow mesenchymal stem cell ameliorates the autoimmune pathogenesis in MRL/lpr mice. Cell Mol Immunol 5:417–424

Chapter 8
Stem Cells Therapy for Cardiomyopathy: An Emerging Paradigm

Kanwal Rehman, Komal Jabeen, and Muhammad Sajid Hamid Akash

8.1 Introduction

Cardiomyopathy (CMP) is a genetic disorder representing a collection of diverse heart muscle's conditions affecting many peoples globally. CMP interprets a major role in health care system owing to increased ratio of morbidity and mortality associated with high risk of heart failure, heart transplantation, cardiac arrhythmias, stroke, and death. More than two decades ago, the first genetic cause of CMP was invented (Geisterfer-Lowrance et al. 1990). Genetic disorder is basically the abnormality in one or more genome that may be due to the heredity and passed down from parent's gene to offspring. The American Heart Association (AHA) classified CMP into primary and secondary CMPs (Maron 2008; Maron et al. 2006). It was also classified according to the morphology and pathophysiology by the European Society of Cardiology (ESC). Genetic studies play an important role to understand the different types of CMPs. Primary CMP can be genetic, i.e., inherited or acquired. Further classification is based on the functional and anatomical impairment leading to the hypertrophic CMP, dilated cardiomyopathy (DCMP) , restrictive cardiomyopathy (RCMP) , and arrhythmogenic right ventricular CMP. In case of secondary cardiomyopathy (SCMP), systemic diseases are studied that are associated with inflammation, metabolic abnormalities, mitochondrial disorders, muscular dystrophies and chemical toxin among the others (Elliott et al. 2007).

K. Rehman · K. Jabeen
Institute of Pharmacy, Physiology and Pharmacology, University of Agriculture, Faisalabad, Punjab, Pakistan
e-mail: kanwal.akash@uaf.edu.pk

M. S. H. Akash (✉)
Department of Pharmaceutical Chemistry, Government College University Faisalabad, Faisalabad, Punjab, Pakistan
e-mail: sajidakash@gcuf.edu.pk

P. V. Pham, A. El-Hashash (eds.), *Stem Cells for Cancer and Genetic Disease Treatment*, Stem Cells in Clinical Applications, https://doi.org/10.1007/978-3-319-98065-2_8

If CMP often goes undiagnosed, number of people suffering from this disease varies, i.e., 1 out of 500 (that's about 600,000 Americans) has suffered this condition. CMP may affect the males and females of all age groups and races (Maron 2008). More than two million have been affected from CMP and myocarditis in 2015 ("Global, regional, and national incidence, prevalence, and years lived with disability for 310 diseases and injuries, 1990–2015: a systematic analysis for the Global Burden of Disease Study 2015," 2016). Replacement of CMPs and restoration of blood flow is necessary for reversal of CMP, particularly if it is being passed on to the generations as a hereditary disease. Heart has a defined capacity for regeneration; however, cardiac stem cells (CSCs) have been recognized for their ability to repair and/or regenerate cardiomyocytes. CSCs are known to permit regeneration of cardiac muscles and having ability to differentiate, rapidly proliferate, and undergo self-renewal (Fukuda 2003).

Different therapies are being implemented to treat cardiac diseases but stem cells therapy (SCT) is found to be an ideal treatment for myocardial cell regeneration. Other therapies like pharmacotherapy and coronary artery bypass grafting (CABG) have also enabled the process of cardiac muscle tissues regeneration and pathophysiological changes (Michler 2013). Unlimited sources of stem cells have also been identified for cardiac repairment including CSCs, embryonic stem cell (ESCs), somatic stem cells (SSCs), mesenchymal stem cells (MSCs), and hematopoietic stem cells (HSCs). All the mentioned sources have been found to improve and regenerate the cardiomyocytes and myocardium. Types of cells, doses, isolation procedures of cells, and timing of cell transplantation are considered as key factors in determining cardiac functional retrieval (Laflamme and Murry 2011; Segers and Lee 2008; Tongers et al. 2011). Hence, to reduce risk of heart transplantation and other therapies, restore cardiac function and structure, stem cell therapy is to be employed and manipulated with further improvements and details. Hence, this chapter will include the detailed discussion about some major sources and types of stem cells along with their abilities to proliferate and contribute in the regeneration of cardiomyocytes with special reference to genetic disorders like CMP.

8.2 What Is Cardiomyopathy

CMP is a heterogeneous group of diseases, manifested by structural and functional abnormality of myocardium associated with rigidity and enlargement of cardiac muscles resulting in inability of heart to pump the blood efficiently. It may result in irregular cardiac rhythm, heart failure, heart valve problem and in some cases sudden death (Richard et al. 2006). CMP can affect the cardiac muscles and is considered as hereditary (Laflamme and Murry 2011). Occurrence of CMP results in thickening, stiffening, and/or thinning of heart muscles. CMP is also manifested by cardiac dysfunction associated with rigidity and enlargement of cardiac muscles resulting in inability of heart to pump the blood efficiently. It alters the normal cardiac rhythm that may lead to heart failure and sudden death in some cases (Richard et al. 2006). These conditions adversely affect the heart's ability to pump the blood leading to irregular heartbeats and heart failure.

8.2.1 *Types of Cardiomyopathy*

CMP has been classified into different types by AHA and ESC. ESC classified it according to the morphology and pathophysiology, and AHA classified CMP into primary and secondary CMPs. Primary CMP can be genetic, i.e., inherited or acquired. Systemic diseases that are associated with inflammation, metabolic abnormalities, mitochondrial disorders, muscular dystrophies, and chemical toxin are considered as secondary CMP. Primary CMP is further classified based on the functional and anatomical impairment leading to hypertrophic cardiomyopathy (HCMP), dilated cardiomyopathy (DCMP), restrictive cardiomyopathy (RCMP), ischemic cardiomyopathy (ICMP), and arrhythmogenic right ventricular cardiomyopathy (ARV-CMP) (Elliott et al. 2007; Jefferies and Towbin 2010). The prevalence of CMPs differs according to its different types. The incidence of DCMP is more than 50%, occurrence of HCMP is more than 40%, and RCMP and arrhythmic right ventricular cardiomyopathy account for less than 5% of the cases (Lipshultz et al. 2003).

8.2.1.1 Dilated Cardiomyopathy

DCMP is a common CMP worldwide and is a major cause of myocardial disease in which diminished strength of contractility and dilation of left ventricle occur. Impaired left ventricle systolic function and secondary diastolic dysfunction happen with an unknown reason. This condition is called as idiopathic DCMP. It reduces the ratio of morbidity and mortality by reversing the process. Congestive heart failure, valvular defects, conduction impairment, and sudden death are common features of DCMP (Kaur et al. 2013; Sugrue et al. 1992). In some cases, genetic factors are involved in the development of aforementioned disease. Genetic DCMP can be transmitted by autosomal dominant inheritance pattern and/or in rare cases with autosomal recessive or x linked inheritance (Hershberger and Siegfried 2011). In familial DCMP, almost 40 genes are involved and mutation in some genes has also been observed (Millat et al. 2011; Spaendonck-Zwarts et al. 2013; Teekakirikul et al. 2013). Some non-genetic factors are also linked for the development of DCMP including metabolic, endocrine and nutritional disorders, viral and bacterial infections, autosomal disease, and exposure to toxins, i.e., alcohol (Petretta et al. 2011).

8.2.1.2 Hypertrophic Cardiomyopathy

HCMP is a disease of functional impairment of heart muscle that is developed by hypertrophy (enlargement) of a part of myocardium and excessive thickness of the heart muscles (Maron 2002; Maron et al. 2003; Sherrid et al. 2003). The prevalence of HCMP is ultimately 1 in 500 general populations; it is the most common inherited cardiac disease (Maron 2004; Maron et al. 1995). HCMP is caused by the mutation of encoding component of protein that is sarcomere. The mutation pattern is inherited in an autosomal dominant (Alcalai et al. 2008). HCMP patients having sarcomeric mutations are more vulnerable to develop ventricular arrhythmias, heart failure, and cardiovascular death than the patients without mutations (Girolami et al. 2010).

8.2.1.3 Ischemic Cardiomyopathy

In ICMP, the supply of blood to the heart is not appropriate because of narrowing of the coronary artery that is responsible for pumping blood to the heart. The patients previously suffered in acute myocardial infarction are more predicated for ICMP (Felker et al. 2002). The ejection fraction of left ventricle is <35–40% due to impaired ventricle function due to the consequences from coronary artery disease that lead to ICMP (Felker et al. 2002). Progression of ventricle remodeling, hypertrophy, apoptosis and necrosis of cardiomyocytes, inappropriate supply of blood, and poor contractility of left ventricle are caused by ischemia (Pangonytė et al. 2008; Zamilpa et al. 2014).

8.2.1.4 Arrhythmogenic Right Ventricular Cardiomyopathy

ARV-CMP is a genetic disorder characterized by non-ischemic ventricular arrhythmia that mainly affects the right ventricle and in some cases, the involvement of left ventricle is more than 70%. It is one of the major risks of ventricular arrhythmia, biventricular heart failure, and sudden cardiac death (Blomström-Lundqvist et al. 1987; Fontaine et al. 1992; Manyari et al. 1983; Pinamonti et al. 1992; Webb et al. 1986). The estimated risk of ARV-CMP in population is 1 in 5000 approximately (Norman and McKenna 1999). Different causes are responsible for eradication of ARV-CMP that may be genetic, congenital, and acquired factors. The mutation occurs in coding protein holding the cardiac muscle cells. This is a risk produced by genetic factors, thus showing variability in expression due to the transmission of dominant autosomal inheritance pattern (McKenna et al. 1994; Miani et al. 1993). Acquired factor is also responsible for inducing arrhythmogenic dysplasia like viral or inflammatory myocarditis (Calabrese et al. 2006).

8.2.1.5 Restrictive Cardiomyopathy

The least common cardiomyopathy disorder is RCMP. In RCMP, the filling and stretching of ventricles is impaired resulting in dysfunction of diastolic filling but systolic function is preserved and stiffness of ventricles is also compromised (Abelmann 1984). These changes may be resulted from the primary etiology on genetic bases. Systemic disorders like amyloidosis, and idiopathic and storage diseases, i.e., glycogen storage disease, are the secondary etiologies that are responsible to induce RCMP (Topol and Califf 2007).

8.2.2 *Causes of Cardiomyopathy*

All types of CMP indicate an underlying cause and/or reason according to their previous history at the end of clinical evaluation (Felker et al. 2002; Kasper et al. 1994). The possible causes of CMP are as follows:

- Genetic conditions
- Heart failure syndrome
- Toxin (alcohol)
- Long-term high blood pressure
- Connective tissue disorders
- Chronic rapid heart rate
- Drugs notably cocaine, amphetamines, anthracyclines (Billingham 1985; Mason et al. 1977)
- Nutritional deficiencies (Watkins et al. 2011)

Hemochromatosis is an autosomal recessive characteristic in which the iron is built up in the muscle of the heart. Sarcoidosis is a condition in which co-existence of pulmonary disease and characteristics appear on cardiac MI. In this condition, the lumps of cells grow in heart due to inflammation (Bacon et al. 2011). Amyloidosis is a disorder that causes the production of abnormal protein and effect organs like renal, peripheral, connective tissue and skin leading to the progression of cardiac conduction disease (Chew et al. 1975; Cueto-Garcia et al. 1984). Metabolic disorders like obesity, diabetes, and thyroid disease may increase the vulnerability to cardiomyopathy. Use of some chemotherapy drugs and radiation to treat cancer and glycogen storage diseases are also one of the causes leading to CMP (Gilbert 1986; Gottdiener et al. 1983).

8.3 Risk Factors of Cardiomyopathy

8.3.1 *Genetics*

CMP is a genetic disease. Patient having a history of CMP is at higher risk to develop heart failure and sudden cardiac death (SCD) (Watkins et al. 2011).

8.3.2 *Smoking*

In coronary artery surgery study and study of left ventricular dysfunction, more than 40% cases of developing cardiac failure are associated with smoking (Suskin et al. 2001).

8.3.3 Illicit Drug Use

Drugs such as cocaine, amphetamines, and anabolic steroids may also increase the risk of CMP.

8.3.4 Diabetes

Diabetic patients are at high risk of developing CMP, along with heart failure and other cardiac issues.

8.3.5 Miscellaneous Diseases

Some of the diseases which can damage the heart and responsible for the risk of CMP include:

- **Hemochromatosis**: Iron is stored in the body which increases the risk of CMP.
- **Sarcoidosis**: A disease in which lumps of cells grow in the heart and cause inflammation.
- **Amyloidosis**: Building up of abnormal protein and connective tissue disorder can increase the risk of CMP (Bacon et al. 2011).

Some of the other important factors participating in CMP development are enlisted below:

- Long-term high blood pressure
- Viral infections
- Thyroid disorder
- Obesity
- Alcoholism
- Metabolic disorders

8.4 Therapies to Combat Cardiomyopathy

The aim of CMP therapies is to improve the cardiac function, relieve symptoms, prevent cardiac remodeling, shorten hospital stay, reverse injury process, and decrease mortality. A class of drugs is exploited for the management of CMPs. In 2016, The AHA and American college of cardiology recognized a new pharmacological therapy for patients with reduced ejection fraction along with the complications of CMPs and heart failure to reduce morbidity and mortality (Yancy et al. 2016). Following are some therapies which are commonly being used to treat CMP:

- Angiotensin-converting enzyme (ACE) inhibitors
- Angiotensin II receptor blockers (ARBs)
- Beta-blockers
- Aldosterone antagonists
- Cardiac glycosides
- Diuretics
- Nitrates
- Vasodilators
- Antiarrhythmics
- Inotropic agents
- Anticoagulants

Various surgical procedures are also used which indicate no beneficial improvement to pharmacological therapy. A number of patients move to the end stage, e.g., heart failure and sudden cardiac death. Nowadays, SCT is being investigated that prevents the cardiac remodeling and provides a safe and effective treatment because of its renewal properties that can activate the CSC and repair cardiac injury (Blau et al. 2001; Smith et al. 2007).

8.4.1 Stem Cells Therapy

Stem cells are undifferentiated biological cells capable of differentiating and dividing into more stem cells from a parent cell with renewal properties (Blau et al. 2001; Tuch 2006). These play important role in the improvement of cardiac function, myocardial contraction and repair damaged heart by inhibiting cells fibrosis and apoptosis. Stem cells are capable of promoting angiogenesis, proliferation of endogenous cardiac stem cells and differentiation to cardiac muscle cells (Gnecchi et al. 2012). The common treatment for cardiac repairing includes pharmacotherapy, angiography and CABG or percutaneous coronary intervention. These restore the supply of blood and cardiac function in minor degree, but do not cure the pathophysiological alteration. Therefore SCT is the ideal treatment for regeneration of myocardial cells (Michler 2013).

8.4.2 Types of Stem Cells Therapy for the Treatment of Cardiomyopathy

Unlimited sources of stem cells have been identified for cardiac repairment and cardiomyocytes regeneration. Some of the most important sources are given below as described by Abbate et al. (2008).

- Cardiac stem cells
- Mesenchymal stem cells
- Embryonic stem cells
- Bone marrow-derived stem cells

8.4.2.1 Cardiac Stem Cells

Nadal-Ginard and his colleagues in 2003 (Nadal-Ginard et al. 2003) discovered cardiac stem cells (CSCs). CSCs have potential to produce cardiomyocytes, endothelial cells, smooth muscle cells, and regenerative capacity for adult's heart. Stem cells have also self-renewing characteristics (Urbanek et al. 2006). After cardiac injury, loss of structure and function of cardiac muscle cells occur. Newly formed myocytes and cardiac cells are of insufficient capacity for regeneration. CSCs transplantation boosts the heart's own regeneration capacity (Bolli et al. 2011; Urbanek et al. 2006). CSCs are known to permit regeneration of cardiac muscle and have ability to differentiate, rapidly proliferate, and undergo self-renewal. The use of CSCs has attained considerable attraction due to their ability to replicate and recruit other stem cells for the regeneration of cardiac tissue. CSCs are used to treat cardiac disorder and to replace the necrotic tissue with functional myocardium. It is then transported into structural and functional muscle obstruction and/or disorder (Bearzi et al. 2007; Beltrami et al. 2003).

8.4.2.2 Mesenchymal Stem Cells

Young and colleagues described the populations of mesenchymal stem cells (MSCs) that can be derived from connective tissue (Young et al. 1999, 2001b). MSCs have properties of self-renewing and differentiation and used in regeneration of cardiac repairment (Young et al. 2001a). They are isolated from the bone marrow, adipose tissue, and umbilical cord, and share the similar immunomodulatory and cardiopoietic properties (Selem et al. 2011). Survival rate and movement towards the targeted site are central elements to determine the therapeutic effects of MSCs. MSCs secrete some growth factors for cardiac repairment after stimulations which can also be differentiated into the cardiomyocytes. Mostly, ischemic cardiomyopathy, reduction in malignant ventricular arrhythmias, improved pulmonary function, and acute myocardial infarction with percutaneous coronary intervention patients show best therapeutic response after MSCs therapy and left ventricle ejection fraction (LVEF).

8.4.2.3 Embryonic Stem Cells

Embryonic stem cells (ESCs) can produce cardiomyocytes in vivo and in vitro, but can also differentiate into cardiomyocytes (Hescheler et al. 1999; Klug et al. 1996). ESCs are capable of developing specific cells types in vitro like atrial-like, ventricular-like, and sinus nodal-like in the heart (Boheler et al. 2005). Embryonic stem cells are detached from inner mass of human blastocysts and these are pluripotent stem cells (Thomson et al. 1998). In 1998, ESCs were first isolated from the embryos in the inner side of the cell mass. It provided an unlimited source of cells holding a delineate capacity of differentiation and proliferation under defined conditions (Pera et al. 2000). ESCs are helpful in regenerating damaged myocardium and provided versatile types of cells including cardiomyocytes, fibroblast, smooth muscle cells, and endothelial cells. Human ESCs regenerate a large quantity of cardiac stem cells that can differentiate into endothelial and smooth muscle cells in vitro (Vittet et al. 1996).

8.4.2.4 Bone Marrow-Derived Stem Cells

Bone marrow-derived stem cells originate from various types of cells including hematopoietic stem cells (HSCs), MSCs, and endothelial stem/progenitor cells (EPSc). Bone marrow cells are stem cells that can be drawn out from the bone marrow of patients (Dawn and Bolli 2005). Improved cardiac remodeling through the injection of bone marrow-derived stem cells can be differentiated into cardiac components and meliorate the curing of injured myocardium. Bone marrow cells are easily affordable source of cells (Orlic et al. 2001). Bone marrow stem cells are responsible for the origination of cardiomyocytes and by expression of transdifferentiation and cells fusion also impart to the renewal of myocytes after injury (Orlic et al. 2001; Wu et al. 2015). The delivery of autologous bone marrow stem cells is responsible for regeneration of damaged myocardium and scarred area in the heart as investigated by Orlic et al. in 2001 (Orlic et al. 2001).

8.4.3 Advantage of Stem Cells Therapy

CMP and ischemic heart disease are major cardiac issues worldwide in which the abnormality occurs in the structure and function of cardiac muscles. Patients suffering from chest pain, shortness of breath, and limited exercise capacity resulted from unlimited supply of oxygenated blood and cardiac nutrients and cardiomyocytes ischemic injury (De Zwaan et al. 2001). Severe conditions lead to cell necrosis and loss of cardiomyocytes. Lifestyle modifications, pharmacological management, and surgical intervention are available for curing the cardiac problems but a larger population can lead to heart failure. In this scenario, stem cells therapy emerged as a potential strategy for treating CMP and other cardiac related problems (O'Gara et al. 2012). Pharmacotherapy, angiography, and coronary artery bypass grafting (CABG) restore the supply of blood and cardiac function in minor degree but not cure the pathophysiological alteration. Therefore SCT is safe and ideal treatment for regeneration of myocardial cells (Michler 2013).

Stem cells possess self-renewal, proliferation, and differentiation properties (Blau et al. 2001). CSCs stimulate specific cardiac biomarker in the heart that can differentiate into cardiomyocytes and improve cardiac regeneration ability (Carpenter et al. 2011). The goal of stem cells therapy is to repair the injured heart and improve cardiac function in CMP which includes release of factors that can devalue the apoptosis of CMP, provoke angiogenesis in ischemic CMP and other ischemic heart diseases because it effectively improves the myocardial blood flow, genes related to promotion of angiogenesis, i.e., Ang-1, FGF-2, hepatocyte growth factor (HGF), and activation of residential stem cells (Chien et al. 2008; Fuchs et al. 2001; Yang et al. 2007).

8.5 Conclusion

In this chapter, we have briefly discussed the pathophysiology of CMP and its different types that can be considered as major cause of morbidity and mortality globally. Nowadays, new modality for cardiac repairing is the use of stem cells therapy as a remedy of CMP. Several SCTs have been identified for attenuating remodeling to regenerate damaged cardiomyocytes and to improve cardiac function and exercise tolerance. SCT is a safe treatment and improves the quality of life in cardiomyopathic patients. Further studies are needed to investigate the underlying mechanism for better improvement in the use of SCT that can further improve the patient's outcome by using appropriate strategies.

References

Abbate A, Biondi-Zoccai GG, Appleton DL, Erne P, Schoenenberger AW, Lipinski MJ et al (2008) Survival and cardiac remodeling benefits in patients undergoing late percutaneous coronary intervention of the infarct-related artery: evidence from a meta-analysis of randomized controlled trials. J Am Coll Cardiol 51(9):956–964

Abelmann WH (1984) Classification and natural history of primary myocardial disease. Prog Cardiovasc Dis 27(2):73–94

Alcalai R, Seidman JG, Seidman CE (2008) Genetic basis of hypertrophic cardiomyopathy: from bench to the clinics. J Cardiovasc Electrophysiol 19(1):104–110

Bacon BR, Adams PC, Kowdley KV, Powell LW, Tavill AS (2011) Diagnosis and management of hemochromatosis: 2011 practice guideline by the American Association for the Study of Liver Diseases. Hepatology 54(1):328–343

Bearzi C, Rota M, Hosoda T, Tillmanns J, Nascimbene A, De Angelis A et al (2007) Human cardiac stem cells. Proc Natl Acad Sci U S A 104(35):14068–14073. https://doi.org/10.1073/pnas.0706760104

Beltrami AP, Barlucchi L, Torella D, Baker M, Limana F, Chimenti S et al (2003) Adult cardiac stem cells are multipotent and support myocardial regeneration. Cell 114(6):763–776

Billingham ME (1985) Pharmacotoxic myocardial disease: an endomyocardial study. Heart Vessel 1:278–282

Blau HM, Brazelton TR, Weimann JM (2001) The evolving concept of a stem cell: entity or function? Cell 105(7):829–841

Blomström-Lundqvist C, Sabel K-G, Olsson SB (1987) A long term follow up of 15 patients with arrhythmogenic right ventricular dysplasia. Heart 58(5):477–488

Boheler KR, Crider DG, Tarasova Y, Maltsev VA (2005) Cardiomyocytes derived from embryonic stem cells. Methods Mol Med 108:417–435

Bolli R, Chugh AR, D'Amario D, Loughran JH, Stoddard MF, Ikram S et al (2011) Cardiac stem cells in patients with ischaemic cardiomyopathy (SCIPIO): initial results of a randomised phase 1 trial. Lancet 378(9806):1847–1857

Calabrese F, Basso C, Carturan E, Valente M, Thiene G (2006) Arrhythmogenic right ventricular cardiomyopathy/dysplasia: is there a role for viruses? Cardiovasc Pathol 15(1):11–17

Carpenter L, Carr C, Yang CT, Stuckey DJ, Clarke K, Watt SM (2011) Efficient differentiation of human induced pluripotent stem cells generates cardiac cells that provide protection following myocardial infarction in the rat. Stem Cells Dev 21(6):977–986

Chew C, Ziady GM, Raphael MJ, Oakley CM (1975) The functional defect in amyloid heart disease: the "stiff heart" syndrome. Am J Cardiol 36(4):438–444

Chien KR, Domian IJ, Parker KK (2008) Cardiogenesis and the complex biology of regenerative cardiovascular medicine. Science 322(5907):1494–1497

Cueto-Garcia L, Tajik AJ, Kyle RA, Edwards WD, Greipp PR, Callahan JA et al (1984) Serial echocardiographic observations in patients with primary systemic amyloidosis: an introduction to the concept of early (asymptomatic) amyloid infiltration of the heart. Paper presented at the Mayo Clinic Proceedings

Dawn B, Bolli R (2005) Adult bone marrow–derived cells: regenerative potential, plasticity, and tissue commitment. Basic Res Cardiol 100(6):494–503

De Zwaan C, Daemen M, Hermens WT (2001) Mechanisms of cell death in acute myocardial infarction: pathophysiological implications for treatment. Neth Hear J 9(1):30

Elliott P, Andersson B, Arbustini E, Bilinska Z, Cecchi F, Charron P et al (2007) Classification of the cardiomyopathies: a position statement from the European Society Of Cardiology Working Group on Myocardial and Pericardial Diseases. Eur Heart J 29(2):270–276

Felker GM, Shaw LK, O'Connor CM (2002) A standardized definition of ischemic cardiomyopathy for use in clinical research. J Am Coll Cardiol 39(2):210–218

Fontaine G, Frank R, Fontaliran F, Lascault G, Tonet J (1992) Right ventricular tachycardias. In: Parmley WW, Chatterjee K (eds) Cardiology. JB Lippincott Co, New York, pp 1–17

Fuchs S, Baffour R, Zhou YF, Shou M, Pierre A, Tio FO et al (2001) Transendocardial delivery of autologous bone marrow enhances collateral perfusion and regional function in pigs with chronic experimental myocardial ischemia. J Am Coll Cardiol 37(6):1726–1732

Fukuda K (2003) Use of adult marrow mesenchymal stem cells for regeneration of cardiomyocytes. Bone Marrow Transplant 32(S1):S25–S27

Geisterfer-Lowrance AA, Kass S, Tanigawa G, Vosberg H-P, McKenna W, Seidman CE, Seidman J (1990) A molecular basis for familial hypertrophic cardiomyopathy: a β cardiac myosin heavy chain gene missense mutation. Cell 62(5):999–1006

Gilbert E (1986) The effects of metabolic diseases on the cardiovascular system. Am J Cardiovasc Pathol 1(2):189–213

Girolami F, Ho CY, Semsarian C, Baldi M, Will ML, Baldini K et al (2010) Clinical features and outcome of hypertrophic cardiomyopathy associated with triple sarcomere protein gene mutations. J Am Coll Cardiol 55(14):1444–1453

Global, regional, and national incidence, prevalence, and years lived with disability for 310 diseases and injuries, 1990-2015: a systematic analysis for the Global Burden of Disease Study 2015 (2016) Lancet 388(10053):1545–1602. https://doi.org/10.1016/s0140-6736(16)31678-6

Gnecchi M, Danieli P, Cervio E (2012) Mesenchymal stem cell therapy for heart disease. Vasc Pharmacol 57(1):48–55. https://doi.org/10.1016/j.vph.2012.04.002

Gottdiener JS, Katin MJ, Borer JS, Bacharach SL, Green MV (1983) Late cardiac effects of therapeutic mediastinal irradiation: assessment by echocardiography and radionuclide angiography. N Engl J Med 308(10):569–572

Hershberger RE, Siegfried JD (2011) Update 2011: clinical and genetic issues in familial dilated cardiomyopathy. J Am Coll Cardiol 57(16):1641–1649

Hescheler J, Fleischmann B, Wartenberg M, Bloch W, Kolossov E, Ji G et al (1999) Establishment of ionic channels and signalling cascades in the embryonic stem cell-derived primitive endoderm and cardiovascular system. Cells Tissues Organs 165(3-4):153–164

Jefferies JL, Towbin JA (2010) Dilated cardiomyopathy. Lancet 375(9716):752–762. https://doi.org/10.1016/s0140-6736(09)62023-7

Kasper EK, Agema WR, Hutchins GM, Deckers JW, Hare JM, Baughman KL (1994) The causes of dilated cardiomyopathy: a clinicopathologic review of 673 consecutive patients. J Am Coll Cardiol 23(3):586–590

Kaur H, Khetarpal R, Aggarwal S (2013) Dilated cardiomyopathy: an anaesthetic challenge. J Clin Diagn Res 7(6):1174

Klug MG, Soonpaa MH, Koh GY, Field LJ (1996) Genetically selected cardiomyocytes from differentiating embronic stem cells form stable intracardiac grafts. J Clin Investig 98(1):216
Laflamme MA, Murry CE (2011) Heart regeneration. Nature 473(7347):326–335. https://doi.org/10.1038/nature10147
Lipshultz SE, Sleeper LA, Towbin JA, Lowe AM, Orav EJ, Cox GF et al (2003) The incidence of pediatric cardiomyopathy in two regions of the United States. N Engl J Med 348(17):1647–1655. https://doi.org/10.1056/NEJMoa021715
Manyari DE, Klein GJ, Gulamhusein S, Boughner D, Guiraudon GM, Wyse G et al (1983) Arrhythmogenic right ventricular dysplasia: a generalized cardiomyopathy? Circulation 68(2):251–257
Maron BJ (2002) Hypertrophic cardiomyopathy: a systematic review. JAMA 287(10):1308–1320
Maron BJ (2004) Hypertrophic cardiomyopathy: an important global disease. Am J Med 116(1):63–65
Maron BJ (2008) The 2006 American Heart Association classification of cardiomyopathies is the gold standard. Circ Heart Fail 1(1):72–76
Maron BJ, Gardin JM, Flack JM, Gidding SS, Kurosaki TT, Bild DE (1995) Prevalence of hypertrophic cardiomyopathy in a general population of young adults. Echocardiographic analysis of 4111 subjects in the CARDIA Study. Coronary Artery Risk Development in (Young) Adults. Circulation 92(4):785–789
Maron BJ, McKenna WJ, Danielson GK, Kappenberger LJ, Kuhn HJ, Seidman CE et al (2003) American College of Cardiology/European Society of Cardiology clinical expert consensus document on hypertrophic cardiomyopathy: a report of the American College of Cardiology foundation task force on clinical expert consensus documents and the European Society of Cardiology committee for practice guidelines. Eur Heart J 24(21):1965–1991
Maron BJ, Towbin JA, Thiene G, Antzelevitch C, Corrado D, Arnett D et al (2006) Contemporary definitions and classification of the cardiomyopathies: an American Heart Association Scientific Statement from the Council on Clinical Cardiology, Heart Failure and Transplantation Committee; Quality of Care and Outcomes Research and Functional Genomics and Translational Biology Interdisciplinary Working Groups; and Council on Epidemiology and Prevention. Circulation 113(14):1807–1816. https://doi.org/10.1161/circulationaha.106.174287
Mason JW, Billingham ME, Friedman JP (1977) Methysergide-induced heart disease: a case of multivalvular and myocardial fibrosis. Circulation 56(5):889–890
McKenna WJ, Thiene G, Nava A, Fontaliran F, Blomstrom-Lundqvist C, Fontaine G, Camerini F (1994) Diagnosis of arrhythmogenic right ventricular dysplasia/cardiomyopathy. Task Force of the Working Group Myocardial and Pericardial Disease of the European Society of Cardiology and of the Scientific Council on Cardiomyopathies of the International Society and Federation of Cardiology. Br Heart J 71(3):215
Miani D, Pinamonti B, Bussani R, Silvestri F, Sinagra G, Camerini F (1993) Right ventricular dysplasia: a clinical and pathological study of two families with left ventricular involvement. Br Heart J 69(2):151–157
Michler RE (2013) Stem cell therapy for heart failure. Methodist Debakey Cardiovasc J 9(4):187–194
Millat G, Bouvagnet P, Chevalier P, Sebbag L, Dulac A, Dauphin C et al (2011) Clinical and mutational spectrum in a cohort of 105 unrelated patients with dilated cardiomyopathy. Eur J Med Genet 54(6):e570–e575
Nadal-Ginard B, Kajstura J, Leri A, Anversa P (2003) Myocyte death, growth, and regeneration in cardiac hypertrophy and failure. Circ Res 92(2):139–150
Norman M, McKenna W (1999) Arrhythmogenic right ventricular cardiomyopathy: perspectives on disease. Z Kardiol 88(8):550–554
O'Gara PT, Kushner FG, Ascheim DD, Casey DE, Chung MK, De Lemos JA et al (2012) 2013 ACCF/AHA guideline for the management of ST-elevation myocardial infarction. Circulation 127(4):e362–e425. https://doi.org/10.1161/CIR.0b013e3182742cf6
Orlic D, Kajstura J, Chimenti S, Limana F, Jakoniuk I, Quaini F et al (2001) Mobilized bone marrow cells repair the infarcted heart, improving function and survival. Proc Natl Acad Sci U S A 98(18):10344–10349

Pangonytė D, Stalioraitytė E, Žiuraitienė R, Kazlauskaitė D, Palubinskienė J, Balnytė I (2008) Cardiomyocyte remodeling in ischemic heart disease. Medicina 44(11):848–854
Pera MF, Reubinoff B, Trounson A (2000) Human embryonic stem cells. J Cell Sci 113(1):5–10
Petretta M, Pirozzi F, Sasso L, Paglia A, Bonaduce D (2011) Review and metaanalysis of the frequency of familial dilated cardiomyopathy. Am J Cardiol 108(8):1171–1176
Pinamonti B, Sinagra G, Salvi A, Di Lenarda A, Morgera T, Silvestri F et al (1992) Left ventricular involvement in right ventricular dysplasia. Am Heart J 123(3):711–724
Richard P, Villard E, Charron P, Isnard R (2006) The genetic bases of cardiomyopathies. J Am Coll Cardiol 48(9):A79–A89
Segers VF, Lee RT (2008) Stem-cell therapy for cardiac disease. Nature 451(7181):937–942
Selem S, Hatzistergos KE, Hare JM (2011) Cardiac stem cells: biology and therapeutic applications. In: Principles of regenerative medicine, Elsevier Inc
Sherrid MV, Chaudhry FA, Swistel DG (2003) Obstructive hypertrophic cardiomyopathy: echocardiography, pathophysiology, and the continuing evolution of surgery for obstruction. Ann Thorac Surg 75(2):620–632
Smith RR, Barile L, Cho HC, Leppo MK, Hare JM, Messina E et al (2007) Regenerative potential of cardiosphere-derived cells expanded from percutaneous endomyocardial biopsy specimens. Circulation 115(7):896–908
Spaendonck-Zwarts KY, Rijsingen IA, Berg MP, Lekanne Deprez RH, Post JG, Mil AM et al (2013) Genetic analysis in 418 index patients with idiopathic dilated cardiomyopathy: overview of 10 years' experience. Eur J Heart Fail 15(6):628–636
Sugrue DD, Rodeheffer RJ, Codd MB, Ballard DJ, Fuster V, Gersh BJ (1992) The clinical course of idiopathic dilated cardiomyopathy: a population-based study. Ann Intern Med 117(2):117–123
Suskin N, Sheth T, Negassa A, Yusuf S (2001) Relationship of current and past smoking to mortality and morbidity in patients with left ventricular dysfunction. J Am Coll Cardiol 37(6):1677–1682
Teekakirikul P, Kelly MA, Rehm HL, Lakdawala NK, Funke BH (2013) Inherited cardiomyopathies: molecular genetics and clinical genetic testing in the postgenomic era. J Mol Diagn 15(2):158–170
Thomson JA, Itskovitz-Eldor J, Shapiro SS, Waknitz MA, Swiergiel JJ, Marshall VS, Jones JM (1998) Embryonic stem cell lines derived from human blastocysts. Science 282(5391):1145–1147
Tongers J, Losordo DW, Landmesser U (2011) Stem and progenitor cell-based therapy in ischaemic heart disease: promise, uncertainties, and challenges. Eur Heart J 32(10):1197–1206
Topol EJ, Califf RM (2007) Textbook of cardiovascular medicine. Lippincott Williams & Wilkins, Philadelphia
Tuch B (2006) Stem cells-a clinical update. Aust Fam Physician 35(9):719
Urbanek K, Cesselli D, Rota M, Nascimbene A, De Angelis A, Hosoda T et al (2006) Stem cell niches in the adult mouse heart. Proc Natl Acad Sci U S A 103(24):9226–9231
Vittet D, Prandini M-H, Berthier R, Schweitzer A, Martin-Sisteron H, Uzan G, Dejana E (1996) Embryonic stem cells differentiate in vitro to endothelial cells through successive maturation steps. Blood 88(9):3424–3431
Watkins H, Ashrafian H, Redwood C (2011) Inherited cardiomyopathies. N Engl J Med 364(17):1643–1656. https://doi.org/10.1056/NEJMra0902923
Webb JG, Kerr CR, Huckell VF, Mizgala HF, Ricci DR (1986) Left ventricular abnormalities in arrhythmogenic right ventricular dysplasia. Am J Cardiol 58(6):568–570
Wu JM, Hsueh Y-C, Ch'ang H-J, Luo C-Y, Wu L-W, Nakauchi H, Hsieh PC (2015) Circulating cells contribute to cardiomyocyte regeneration after injury: novelty and significance. Circ Res 116(4):633–641
Yancy CW, Jessup M, Bozkurt B, Butler J, Casey DE, Colvin MM et al (2016) 2016 ACC/AHA/HFSA focused update on new pharmacological therapy for heart failure: an update of the 2013 ACCF/AHA guideline for the management of heart failure. J Am Coll Cardiol 68(13):1476–1488
Yang J, Zhou W, Zheng W, Ma Y, Lin L, Tang T et al (2007) Effects of myocardial transplantation of marrow mesenchymal stem cells transfected with vascular endothelial growth factor for the improvement of heart function and angiogenesis after myocardial infarction. Cardiology 107(1):17–29. https://doi.org/10.1159/000093609

Young HE, Steele TA, Bray RA, Detmer K, Blake LW, Lucas PW, Black AC Jr (1999) Human pluripotent and progenitor cells display cell surface cluster differentiation markers CD10, CD13, CD56, and MHC class-I. Proc Soc Exp Biol Med 221(1):63–72

Young HE, Duplaa C, Young TM, Floyd JA, Reeves ML, Davis KH et al (2001a) Clonogenic analysis reveals reserve stem cells in postnatal mammals: I. Pluripotent mesenchymal stem cells. Anat Rec 263(4):350–360

Young HE, Steele TA, Bray RA, Hudson J, Floyd JA, Hawkins K et al (2001b) Human reserve pluripotent mesenchymal stem cells are present in the connective tissues of skeletal muscle and dermis derived from fetal, adult, and geriatric donors. Anat Rec 264(1):51–62

Zamilpa R, Navarro MM, Flores I, Griffey S (2014) Stem cell mechanisms during left ventricular remodeling post-myocardial infarction: repair and regeneration. World J Cardiol 6(7):610–620. https://doi.org/10.4330/wjc.v6.i7.610

Chapter 9
Stem Cell Roles and Applications in Genetic Neurodegenerative Diseases

John Ku and Ahmed El-Hashash

Abbreviations

AD	Alzheimer's disease
BDNF	Brain-derived neurotrophic factor
ESC	Embryonic stem cell
HSC	Hematopoietic stem cell
MSC	Mesenchymal stem cell
NSC	Neural stem cell
PD	Parkinson's disease
UCB-MSCs	Umbilical cord blood-derived MSCs

9.1 Introduction

In this era of therapeutic research, the regenerative medicine has emerged to the top of this field since its goals are the restoring, maintaining, and improving of the body functions (Polak et al. 2008). The regenerative medicine also aims to restore the impaired functions by repairing cells, tissues, and organs (Daar and Greenwood 2007). The human body possesses a system to regenerate and repair itself using stem cells in all tissue types. Stem cells can play important roles in both regenerative and translational medicine.

The neurodegeneration describes the progressive decline in the neuronal functions or structures, which can lead to cell death (Rubinsztein 2006). Alzheimer's,

J. Ku
California State Polytechnic University, Pomona, California, USA

A. El-Hashash (✉)
The University of Edinburgh-Zhejiang International campus (UoE-ZJU Institute), and Centre of Stem Cell and Regenerative Medicine, Schools of Medicine & Basic Medicine, Zhejiang University, Haining, Zhejiang, China

P. V. Pham, A. El-Hashash (eds.), *Stem Cells for Cancer and Genetic Disease Treatment*, Stem Cells in Clinical Applications, https://doi.org/10.1007/978-3-319-98065-2_9

Parkinson's, Huntington's, and amyotrophic lateral sclerosis are neurodegenerative diseases that possess distinct morphological and pathophysiological characteristics. These neurodegenerative diseases have common neuropathological features, including some abnormalities in protein functions, faulty protein aggregation and degradation, oxidative stress, and mitochondrial dysfunctions, as well as neuro-inflammation, leading to a progressive loss of brain cell populations (Jellinger 2010; Cacciatore et al. 2012). The loss of neurons that are called the "nucleus basalis" in the substantia innominata of the basal forebrain in Alzheimer's disease (AD) is associated with an increase in both the acetylcholine (Ach) and choline acetyltransferase (ChAT). Similarly, Parkinson's disease (PD) shows a neuron loss in the substantia nigra, which exists in the midbrain and is responsible for the reward, addiction, and body movement. In addition, other studies show that several factors such as the alterations of energy status of degenerating neurons, ubiquitin-proteasome system defects, presence of abnormal aggregated proteins (e.g., both tau and β-amyloid proteins in AD, and both α-synuclein and parkin proteins in PD), the lack of trophic factors, and changes in cytokine levels, as well as disruptions of both the ionic gradients and signal transduction processes, likely affect the specificity of neurodegenerative processes (Farooqui and Horrocks 2007; Fasano et al. 2008).

Two categories of stem cells have been identified: embryonic stem cells (ESCs) and non-embryonic stem cells (non-ESCs). ESCs are totipotent, a property that allows ESCs to give rise to all the embryonic germ layers. Non-ESCs are multipotent cells and called adult stem cells. They can differentiate into limited types of cells in the body compared to ESCs (Lee and Hui 2006).

Stem cells have two essential characteristics that are the ability to self-renew and differentiate into different cell lineages through specific signaling pathways (Ramakrishna et al. 2011; El-Hashash 2016a, b). In one of the earliest stem cell research studies, Becker and colleagues injected cells from the bone marrow into irradiated mice. They found that each newly formed nodule comes from a single bone marrow cell that may have the capability of an infinite self-renewal (Becker et al. 1963).

The stem cell's potency refers to its differential potentials into other cell types and is a subject of intensive research. The most potent stem cell begins with the sperm-ovum fusion, which produces totipotent stem cells that can give rise to both embryonic and extraembryonic cells. While the human totipotent stem cells can differentiate into different types of cells in the body, the pluripotent stem cells exist in the inner cell mass of the blastocyst (blastula) and can give rise to cells forming the three germ layers. Pluripotent embryonic stem cells can, therefore, produce different cell types in the body, excluding the umbilical cord and placenta.

The multipotent stem cells such as both adult and cord blood stem cells can give rise to many body cell types that are limited compared to pluripotent cells. A hematopoietic stem cell, for example, can differentiate into platelets, and both red and white blood cells. In contrast, unipotent stem cells can only differentiate into one cell type, which still possesses a self-renewal ability that differs from non-stem cells (Avasthi et al. 2008; El-Hashash 2016a, b). The existence of stem cells with different degrees of potency and plasticity is important to tissue repair and regeneration (Singh and Williams 2008). In addition, the development and behavior of stem cells are affected by the epithelial to mesenchymal transitions (EMTs) that is also crucial for both cell migration and maintenance of stemness (Thiery et al. 2009; Li et al. 2014; Forte et al. 2017; Chen et al. 2017). EMTs also affect other biological processes such as cell invasion, cause a reduction in cell apoptosis, and contribute to the immunosuppression (Thiery et al. 2009; Chen et al. 2017). In this book chapters, we will review different types of stem cells and describe their roles and applications in genetic neurodegenerative diseases.

9.2 Stem Cell Types and roles in regenerative Medicine

9.2.1 Embryonic Stem Cells (ESCs)

ESCs were first derived from murine blastocysts in 1981 (Evans and Kaufman 1981; Martin 1981), and later from human blastocysts and fetal tissues in 1998 (Thomson et al. 1998). Human ESCs exist in the embryos before their implantation in the uterus, and can differentiate into all cell types in the body (Bajada et al. 2008). The first differentiation process during embryogenesis occurs when an outer cell layer becomes separated from the inner cell mass (ICM) to form the trophectoderm and placenta (Ramakrishna et al. 2011).

9.2.2 Embryonic Germ Cells (EGCs)

The primordial germ line cells of early fetal tissues give rise to EGCs. Additionally, the potential of EGCs is restricted compared to ESCs, because they exist for a limited time that is usually during 5–9 weeks of the embryonic development (Adewumi et al. 2007). Some studies have focused on characterizing EGCs. For example, Shamblott and colleagues have described the characteristics of the gonadal ridge-derived germ cells in humans (Shamblott et al. 1998).

9.2.3 Amniotic Epithelial Cells

During the embryonic development, the embryo sheds a variety of cells into the amniotic fluid. The amniotic epithelial cells (AECs) in humans, for example, are derived from the amniotic membrane in the placenta and express characteristic markers of both ESCs and EGCs. These markers include the alkaline phosphatase, Nanog, and Oct-4 (the octamer-binding transcription factor 4). AECs can differentiate in culture into cell lineages from all germ layers such as the neural cells from the ectoderm, pancreatic endocrine cells, and cardiomyocytes from the mesoderm as well as hepatocytes from the endoderm (Miki et al. 2006; De Coppi et al. 2007).

9.2.4 *Roles of Embryonic Stem/Progenitor Cells in Regenerative Medicine*

Accumulated evidences have indicated that cardiomyocytes and neural stem/progenitor cells that are derived from human ESCs could successfully be transplanted into the animal model of myocardial injury and Parkinson's disease, respectively (Kofidis et al. 2004; Liu et al. 2008; Singh and Williams 2008). In addition, ESC-derived tissues were previously used in animal models of Parkinson's disease (Frenck et al. 1998), the myocardial infarction (Rufer et al. 1999), and the spinal cord injury (Reyes et al. 2001). Notably, in several neurodegenerative disorders, more success has been achieved when transplanting neural progenitors or neuronal cells that are fully matured.

Several studies on human ESCs have indicated that undifferentiated ESCs express a low level of human leukocyte antigen (HLA), a class I antigen that is increased when cells become mature (Martin et al. 2005). Intriguingly, other studies have demonstrated the irreversible arrest of stem cell division with the embryonic death (Landry and Zucker 2004; Gardner 2007). Since ESCs could produce all germ layers, there is a growing interest in patient-specific ESCs for future clinical applications in the transplantation therapies (Ma et al. 2008).

9.2.5 *Adult Stem/Progenitor Cells*

Adult stem cells are non-ESCs that exist in developed tissues and encompass a variety of undifferentiated cell populations, which act as reservoirs to replace cells during normal organ development or tissue turnover, or as needed. The use of non-ESCs is currently a fast-growing and active area of investigation for cell-based therapies (Ramakrishna et al. 2011). Different types of adult stem cells can be defined by their characteristic high proliferation potential and self-renewing capacity as well as their differentiation ability into a mature and functional progeny (Jones et al. 2010), as exemplified below. However, the potency and proliferative potential of adult stem cells are generally lower than ESCs.

9.2.5.1 Hematopoietic Stem Cells (HSCs)

MSCs and HSCs of the bone marrow have been intensively studied to understand their capabilities (Mezey et al. 2000; Weissman 2000; Gardner 2007; Bajada et al. 2008). In addition, both HSC compartment and niche are well characterized (Abdelkrim et al. 2009). HSCs possess certain morphologies and cell-surface marker types, which make them trackable in blood circulation and, therefore, can be isolated for in culture (Lagasse et al. 2000; Adewumi et al. 2007). HSCs are less than 1% of the total bone marrow. However, they can form all blood lineage types (Abdelkrim et al. 2009). Furthermore, HSCs possess the ability to home into specific niches and to engraft themselves into the bone marrow, which is enhanced by the stabilization of hypoxia-inducible factor 1-α (HIF1α; Speth et al. 2014).

Studies on HSCs show that they can produce all blood cell elements that were successfully used for both bone marrow and stem cell transplantation (Hatzimichael and Tuthill 2010). In addition, some of HSCs have the ability to differentiate outside their tissue of origin, as shown in several reporter assay studies. In addition, while neural tissues may act as a source for HSCs (Jackson et al. 1999; Galli et al. 2000; Ma et al. 2008), the bone marrow stroma containing MSCs may also produce neurons and glia (Liechty et al. 2000; Mezey et al. 2000; Woodbury et al. 2000; Ma et al. 2008; Abdelkrim et al. 2009).

9.2.5.2 Neural Stem and Progenitor Cells (NSCs)

NSCs are self-renewable and can generate cells of the glial and neuronal lineages (Abdelkrim et al. 2009). NSCs were successfully isolated from both adult and embryonic brains. In addition, adult NSCs can differentiate to produce different cell types, including neurons and astrocytes, as well as oligodendrocytes (Ramakrishna et al. 2011). The multipotency of NSCs is evident by their contribution to the formation of multiple tissue types when injected into murine blastocysts (Clarke et al. 2000). Furthermore, NSCs can migrate toward damaged areas and proliferate in the neurogenic regions in a brain-injury model (Brignier and Gewirtz 2010).

9.2.5.3 Mesenchymal Stem Cells (MSCs)

MSCs are a group of stromal cells that can self-renew and differentiate into both embryonic and mesodermal cell lineages. These lineages include adipocytes, osteocytes, chondrocytes, neurons, hepatocytes, and both epithelial and muscle cells (Si et al. 2011). In addition, MSCs have both immunomodulation and anti-inflammatory properties (Nauta and Fibbe 2007; Oh et al. 2008). Furthermore, MSCs have the ability to modulate local environments by secreting various types of molecules and factors that may also activate other progenitor cells (Togel et al. 2007). In addition, MSCs can hypersecrete growth factors by a genetic modification and, therefore, have a great therapeutic potential (van Velthoven et al. 2010a, b). MSCs show proliferative properties in culture, and can maintain their undifferentiated multipotency (da Silva Meirelles et al. 2008).

MSCs can produce many growth factors and cytokines in culture and in vivo. These growth factors and cytokines can promote restorative mechanisms, leading to the recovery of the bodily functions (Lu et al. 2001; Mahmood et al. 2004). Moreover, MSCs can provide a remarkable environmental support to other stem cell types. For example, MSCs can support hematopoietic stem cells through the production of certain ligands (Hu et al. 2016). Interestingly, clinical trials on MSCs have shown promises for the treatment of several disorders/diseases and for tissue regeneration (Si et al. 2011).

MSCs can migrate from the blood to injured and/or inflamed tissues, and cause therapeutic effects in vivo (Chavakis et al. 2008). The mechanisms of MSC migration may involve the upregulation of specific ligands in response to injury that facilitates MSC adhesion, and provides a specialized niche that supports the self-renewal and maintains the multipotency of MSCs (Chapel et al. 2003). In addition, these mechanisms include chemokine, integrins, and selectin receptors that are expressed by MSCs and help in MSC migration across the endothelium. Moreover, MSCs can be arrested in micro-capillaries, and then interact with accessory cells to release some trophic cytokines and growth factors (Le Blanc and Ringden 2007; Cselenyak et al. 2010). Notably, murine MSCs can migrate toward the bleomycin-injured mouse lung, but avoid migration toward healthy cells (Ponte et al. 2007). Furthermore, MSCs can migrate toward various types of growth factors and chemokines in culture. This chemokine-driven migration is dose-dependent, and stimulated by the pro-inflammatory cytokine TNF-α, indicating that chemoattraction can direct MSCs to the sites of inflammation (Ponte et al. 2007).

There are several advantages of MSCs, including their availability and low immunogenicity, over other stem cell types that make them ideal candidates for regenerative medicine applications. The main sources of MSCs for tissue regenerations are the adipose tissue, umbilical cord, and bone marrow (Castillo-Melendez et al. 2013). The autologous origin of MSCs reduces their rejection by the immune system (Brehm et al. 2002). In addition, there is a reduction in the tumorgenesis of MSCs because of being adult cells (Brehm et al. 2002).

9.2.5.4 Bone Marrow-Derived MSCs

Hematopoietic and mesenchymal stem cells have multiple potentials and are the major components of the adult bone marrow. Most data on MSCs come from studies on the bone marrow-derived MSCs (BM-MSCs; de Girolamo et al. 2013). BM-MSCs could be obtained from the bone marrow, and easily expanded ex vivo for various applications of autologous transplantation. BM-MSCs have a reduced risk of rejection by the body because of their allogeneic and non-immunogenic properties (Dharmasaroja 2009). In addition, BM-MSCs could survive for approximately 5 months after their in vivo transplantation (Ye et al. 2007).

Furthermore, BM-MSCs are routinely used as positive controls and compared with MSCs that are isolated from other tissues (de Girolamo et al. 2013). Recent studies show that BM-MSCs have a secretory function, and can secrete a specific microRNA type that is able to repair the acute necrotizing pancreatitis in rats (Qian et al. 2016).

9.2.5.5 Adipose Tissue-Derived MSCs

The adipose tissue is a convenient source for MSCs (Li and Xia 2014). MSCs derived from the adipose tissue (ADMSCs) could be produced in a large quantity and, therefore, are a good MSC source for cell-based therapeutic approaches

(Mizuno 2009). The ease of extraction and high rate of proliferation of ADMSCs make them better than other stem cell types for different stem cell applications (Levi et al. 2010).

9.2.5.6 Umbilical Cord Blood-Derived MSCs (UCB-MSCs)

The UCB-MSCs are not easily obtained compared to other MSCs because of their low isolation rate of less than 30% (Si et al. 2011). In contrast, the umbilical cord blood is easily obtained, and represents an alternative of allogeneic MSCs (Markov et al. 2007).

9.2.5.7 Amniotic Membrane-Derived MSCs

The amniotic membrane is another MSC source (Alviano et al. 2007). The amniotic membrane-derived MSCs have multipotent capacity, anti-inflammatory abilities, and low immunogenicity (Chang et al. 2010).

9.3 MSCs and Neurodegenerative Diseases

MSCs were suggested as a promising and viable treatment for both progressive degenerative diseases and acute injury of the nervous system (Himes et al. 2006), stroke (Park et al. 2008), amyotrophic lateral sclerosis (Choi et al. 2010), autoimmune encephalomyelitis (EAE; Zhang et al. 2006), and multiple system atrophy (MSA; Lee and Park 2009). Remarkably, the direct conversion of MSCs into neuronal cells was recently reported by using a combination of small molecules and growth factors (He et al. 2017). Sox2 plays a critical role in this process since its upregulation can regulate the acquisition of neuronal cell fate by MSCs. (He et al. 2017). MSC neuroprotective effects of MSCs are well documented (Castillo-Melendez et al. 2013).

A cell-based therapeutic approach using MSCs was used in both adult animal models and clinical trials of Huntington's disease and traumatic brain injury (Dharmasaroja 2009). In addition, more research methods have been recently developed to promote the neuronal differentiation of MSCs using amyloid hydrogels, which mimic the natural extracellular matrix and help in MSC delivery and engraftment in a Parkinsonian murine model (Das et al. 2016).

MSCs may promote the tissue recovery through the secretion of some soluble factors, including growth factors and cytokines, and insoluble factors such as the extracellular matrix proteins that enhance the survival and regeneration of neural cells through paracrine signaling pathways (Seo and Cho 2012). Studies have also been done on MSC functions in protective processes such as the re-myelination of the brain (Akiyama et al. 2012). Tissue regeneration is, therefore, promoted by MSCs and their secretions of a range of factors. These factors are found in the MSC microenvironment, and can trigger MSC cell proliferation, migration, differentiation, and survival as well as tissue repair (Caplan and Dennis 2006; Yi et al. 2012). In addition, MSCs can decrease immune reactions, inflammation, and cell apoptosis (Caplan and Dennis 2006; Yi et al. 2012).

The mechanisms that regulate the neural repair by MSCs are currently well investigated. Pre-clinical data show that the transplantation of MSCs can lead to a functional recovery after a traumatic brain injury or a hypoxic ischemic in the adult (Parr et al. 2007). In addition, it is possible to direct MSCs toward both cells of the glial (astrocytic) and neuronal cell lineages in vitro (Woodbury et al. 2000; Deng et al. 2006). Moreover, MSC can travel distances to preferentially differentiate into mature neurons after grafting into the lateral ventricle of a developing mouse brain (Deng et al. 2006). This suggests that MSCs are specifically primed toward the neural fate (Deng et al. 2006). Remarkably, MSCs can also be induced into cells possessing a Schwann cell morphology, and express the S-100 marker in vitro (Zhang et al. 2004). Furthermore, the in vitro differentiation of MSC can produce neuron-like cells, expressing some neural markers, including nestin, and the neuronal-specific nuclear protein (NeuN; Sanchez-Ramos et al. 2000).

MSCs may help in stimulating the proliferation of neural stem/progenitor cells by secreting soluble factors, and increasing GFAP expression levels in vitro (Galindo et al. 2011; Munoz et al. 2012). After the brain ischemia, MSC transplantation leads to increased populations of oligodendrocyte progenitor cells and mature oligodendrocyte cells (van Velthoven et al. 2010a, b). It also stimulates myelin formation in the ischemic hemisphere (van Velthoven et al. 2010a, b). In addition, transplantations of human MSCs into injured monkey penumbra can lead to a decrease of both lesion volume and cell death (Xin et al. 2010). In addition, the administration of human MSCs after brain injury results in an improvement of functional recovery in rats. It also hosts cell survival by increasing the activity of protein kinase B (PKB), but decreasing the cleavage of caspase-3 (Kim et al. 2010). Furthermore, the administration of MSCs into the murine models of stroke can result in an increase in the density of axonal fiber and myelination as well as synaptogenesis (Xin et al. 2010).

One of the important aspects of MSC reparative functions is the ability of MSC to induce certain trophic factors in response to injury (Li et al. 2002). These trophic factors include the brain-derived neurotrophic factor (BDNF), fibroblast growth factor-2, fibronectin, glial cell-derived neurotrophic factor (GDNF), heparin binding-epidermal growth factor-like growth factor (HB-EGF), interleukin-6 (IL-6), leukemia inhibitory factor (LIF), platelet-derived growth factor (PDGF), and vascular endothelial growth factor (VEGF; Li et al. 2002). Other characteristic aspects of MSCs are their immunosuppressive properties, and direct interaction with cells of the central nervous system (CNS), as well as their ability to communicate with immune cells (Li et al. 2002). In addition, MSCs can inhibit both T-cell and B-cell proliferation and suppress the functions of natural killer cells, as well as change microglia and macrophage secretory profiles (Le Blanc 2003). Moreover, MSC treatments can lead to a decrease of T-lymphocyte activity, indicating the immunoregulatory capacity of MSCs (Fibbe et al. 2007).

In the ischemic brain, MSC transplantations can lead to the reduction of inflammatory cells that are activated by both macrophages and microglia (Kim et al. 2008). In addition, MSCs can reduce the inflammatory response by the inhibition of microglial expression, which favors the neuronal generation and integration. After MSC treatments of a traumatic brain injury, MSCs may influence the inflammatory response by altering the level of expression of certain anti- and pro-inflammatory cytokines in rats (Galindo et al. 2011). In addition, the treatment of spinal cord injury using MSCs in rats can result in decreasing the pro-inflammatory cytokine IL-1β and activated microglia, but increasing the anti-inflammatory IL-10 (Seo and Cho 2012). Moreover, another spinal cord injury study showed that MSCs can induce a shift in the phenotype of macrophages from a M1 pro-inflammatory phenotype to a M2 anti-inflammatory state, which can thereby promote both protection and regeneration responses (Busch et al. 2011).

In the injured brain, the tissue regeneration requires both angiogenesis and vascular remodeling. When MSCs were intravenously injected in an animal model of stroke, they migrate toward the ischemic areas to increase angiogenesis and improve the functional recovery of the damaged brain (Li et al. 2002). In addition, these MSCs can produce the required cytokine milieu for promoting the proper angiogenic response (Li et al. 2002). These MSCS can also produce pro-angiogenic factors (e.g., VEGF), and support angiogenesis by regulating the activity of angiogenin and IL-6, as well as growth factors such as basic fibroblast growth factor (bFGF), and bone morphogenetic protein 2 (BMP-2; Li et al. 2002). Interestingly, MSCs may differentiate and give rise to endothelial cells (Potapova et al. 2007).

9.4 Therapeutic Strategies of Neurodegenerative Diseases Using Stem Cells

A cell replacement is a promising therapeutic strategy in the treatment of some types of neurodegenerative diseases (Steiner et al. 2006). Stem cells that are obtained from embryonic or adult tissues can migrate toward damaged regions in lesioned brains of mice or rats. Stem cells engraft themselves in these damaged regions, before proliferation and maturation into functional neurons (Blundell and Shah 2015). In addition, neural precursor cells can migrate into damaged areas to induce a functional recovery after intravenous administrations (Blundell and Shah 2015). Therefore, it has been widely proposed that stem cell implantation can lead to the replacement and/or repair of damaged neuronal cells that are damaged or lost in a variety of diseases, including Alzheimer's disease, Parkinson's disease, Huntington's disease, and amyotrophic lateral sclerosis (Baizabal et al. 2003; Singh and Williams 2008; Ma et al. 2008; Sheyn et al. 2010). We will discuss this process in both Alzheimer's and Parkinson's diseases since they are widely distributed and common age-related neurodegenerative disorders.

9.4.1 Alzheimer's Disease (AD)

AD leads to both memory impairments and dementia in the elderly (Tayeb et al. 2012). This disease was discovered and reported by Alois Alzheimer, a German neuropathologist, in 1906 (Berchtold and Cotman 1998). Alois Alzheimer identified AD in a 50-year-old woman before her death in 1906 (Ulrike and Konrad 2003). AD is more often diagnosed in those who are over 65 years old. However, early-onset Alzheimer's cases may also occur (Mayeux 2003). In 2010, 35.6 million people were estimated to suffer from AD worldwide. One in 85 people is estimated be affected by the year of 2050 worldwide (Rocca et al. 2011; Prince et al. 2013).

The pathological characteristics of Alzheimer's disease include losing neuronal cells, depositing beta-amyloid (Aβ) peptide extracellularly, and formation of neurofibrillary tangles (NFTs) that are composed of hyper-phosphorylated tau, as well as dysfunction of neurotransmitters (Tayeb et al. 2012). In addition, the distribution of amyloid plaques may be seen throughout the neocortex and hippocampus in AD patients (Arnold et al. 1991). The predominate areas of severe neuronal degeneration occur in neuronal cell populations such as the hippocampus, pyramidal neurons of the amygdala, entorhinal cortex, and inferior temporal cortex (Gomez-Isla et al. 1997). Notably, the frontal cortex is protected from the degeneration of pyramidal neurons (Gomez-Isla et al. 1997).

The AD course is apparently unique for everyone, but there are some common symptoms to all individuals. The earliest symptoms may be considered as manifestations of stress or things that are thought as "age-related" concerns (Waldemar et al. 2007). The most common symptom is the difficulty in recalling recent and observed facts, which is due to a reduction or loss of the ability to acquire new memories that may gradually lead to both bodily function loss and death (Tabert et al. 2005). The average life expectancy following a diagnosis of AD can range from 3 to 9 years (Querfurth and LaFerla 2010).

The available treatments for AD, including acetylcholinesterase inhibitors, give a partial and short-term alleviation of symptoms only (Birks 2006; Lanctôt et al. 2009). Another treatment option is the cholinergic neurons transplantation that prevents cognitive decline, which is not a long-lasting approach (Yue and Jing 2015). In addition, the course of Alzheimer's could be modified using stem cells that are genetically modified and can migrate toward target areas (Yue and Jing 2015).

Cell-based therapy has the differentiation potential to form the requisite cell types and cause axonal regeneration, as well as direct synaptogenesis (Yue and Jing 2015). Indeed, stem cells that are engrafted into target areas can form neurons for both signal processing and transmission, and provide oligodendrocytes for re-myelination, as well as produce neurotrophic factors that can protect the host cells (Singh and Williams 2008). Remarkably, the brain-derived neurotrophic factor (BDNF) was recently shown to be critical for both the pathogenesis and treatment of AD (Wu et al. 2016). The BDNF overexpression could improve the therapeutic potentials of engrafted neural stem cells (NSCs) into the damaged areas for AD treatments (Wu et al. 2016). Gain of BDNF function in engrafted NSCs has indeed promoted AD therapeutic potentials (Wu et al. 2016). Moreover, MSCs may enhance the formation of autolysosome that results in the clearance of amyloid-β accumulation, leading to an increase of the neuronal survival against the buildup of Aβ toxicity (Shin et al. 2014).

9.4.2 Parkinson's Disease (PD)

PD is a common neurodegenerative disease that is accompanied by extrapyramidal motor dysfunctions (Olanow et al. 2004). PD is named after James Parkinson, a scientist who discovered and reported the descriptions on shaking palsy in 1817 (Parkinson 2002).

PD is common in people age 50 and above. PD symptoms include resting tremors, bradykinesia (slower and decreased amplitude of movement), and postural stability and rigidity (Chaudhuri and Schapira 2009). The progression of PD can lead to the autonomic failure, depression, dementia, sleep abnormalities, and other symptoms (Chaudhuri and Schapira 2009). In addition, PD is marked by a decline of neuronal cell populations that occurs progressively and includes the neuromelanin-laden dopaminergic neurons of the pars compacta of the substantia nigra, cholinergic nucleus basalis of meynert, aminergic brain stem nuclei, hypothalamic neurons, neurons of the olfactory bulb and sympathetic ganglia, and small cortical neurons, as well as parasympathetic neurons in the gut (Fearnley and Lees 1991). The cell loss pattern is unique in PD and, therefore, is not according to normal aging patterns (Han et al. 2015). In addition, PD can lead to the striatal dopamine loss in both intermediate and dorsal and putamen subdivisions, which may be the cause of the akinesia and rigidity (Kish et al. 1988). Another prominent feature of PD is the existence of neuritis that are degenerated and ubiquitin-positive processes in neuronal cells (Gai et al. 1995). Neuritis is detected in all PD-affected regions of the brain stem (Gai et al. 1995). Other two pathological hallmarks of Parkinson's that are found in the postmortem brain of patients are the existence of both Lewy bodies (proteinaceous inclusion bodies), and reactive microgliosis in the affected areas (Kim et al. 2012).

PD diagnosis is based on clinical manifestations of the disease (Fasano et al. 2008). There is yet to be a reliable biomarker that can detect early neurodegenerations in PD. Shortage of reliable PD biomarkers apparently prevents the detection and monitoring of PD candidate drugs. However, there are some PD biomarker candidates that are promising, including antibodies against neuromelanin, and pathological forms of α-synuclein, as well as DJ-1 (Gerlach et al. 2012).

Two general approaches are currently used for PD treatments with medication. The first approach is based on attempts to increase the dopamine levels in the brain, while the second approach aims to improve the symptoms of the disease by different means (Jankovic and Aguilar 2008). Levodopa (also called L-dopa) is the most widely used treatment for Parkinson's (Jankovic and Aguilar 2008). The dopamine precursor L-3,4-dihydroxyphenylalanine (L-dopa) is normally converted into dopamine through the enzymatic activity of dopa decarboxylase in dopaminergic neurons (Nagal and Singla 2012). Administration of L-dopa temporarily relieves motor symptoms by providing dopamine in the substantia nigra (Kim et al. 2012). However, this dopamine can be also metabolized elsewhere, leading to several side effects, including nausea and dyskinesias, as well as joint stiffness (Kostic et al. 1991). In addition, carbidopa, a peripheral dopa decarboxylase inhibitor, helps to prevent L-dopa metabolism before reaching dopaminergic neurons to reduce side effects. The current and common preparation given to patients is called carbidopa/levodopa (co-careldopa, sinemet®; Abdel-Salam 2008).

A great interest has been instilled in using human ESCs to provide abundant supplies of dopamine neurons for transplantation into PD patients (Thomson et al. 1998). Scientists can produce mature neurons from ESCs in vitro, with up to 30% of these neurons possessing features of dopamine-producing cells (Chuang et al. 2015; Rigamonti et al. 2016; Grade and Götz 2017; Terrigno et al. 2018). These ESC-derived dopamine neurons are without an embryoid body formation (Han et al. 2015).

The clinical trials of human fetal dopaminergic neuron transplantations are promising, leading to remarkable improvements in PD patients (Lindvall et al. 2004). Indeed, cells with dopaminergic neuronal properties were generated in culture from different sources, such as bone marrow stem cells, ESCs, and fetal brain stem cells (Takagi et al. 2005). In addition, ESCs and NSCs, which are derived from either fetal or adult brain, were used to generate dopaminergic neuronal cells (Ho and Li 2006; Sanberg 2007; Singh and Williams 2008).

The current stem cell graft approach begins with pre-differentiation of dopaminergic neurons in culture before transplantation (Han et al. 2015). It can also start with stem cell differentiation in vivo after transplantation into the substantia nigra or striatum (Han et al. 2015).

Finally, it is important to identify new methods that can hinder the progression of PD (Anderson and Lenz 2006; Anderson et al. 2006; Singh and Williams 2008). One way of protection against the neuronal death is to transplant stem cells that express molecules with a neuroprotective property, such as glial-cell-line-derived neurotrophic factor (GDNF; Behrstock et al. 2006). The need of PD patients for these methods remarkably varies, with some patients need several implants in different areas of the brain to reach the optimal recovery (Piccini et al. 2005).

9.5 Conclusions and Future Perspective

Stem cell therapy could provide effective cures for most neurodegenerative diseases. Both human ESCs and adult stem cells can be used for this kind of therapeutic approach. The cell replacement therapy could cure some neurological disorders, including AD and PD. This important function of stem cells is supported by several evidences of stem cell migration toward the damaged brain regions after grafting. Then, they begin to differentiate into functional neurons. In addition, neural precursor cells can be administered intravenously to induce functional recovery in patients. Moreover, more research is still needed for the successful generations of dopamine neurons that will indeed provide an effective and long-lasting treatment for PD patients. The use of stem cells for treating neurodegenerative diseases is the way to face our current "aging era."

References

Abdelkrim H, Juan DB, Jane W, Mohamed A, Bernat S (2009) The immune boundaries for stem cell based therapies: problems and prospective solutions. J Cell Mol Med 13:1464–1475

Abdel-Salam OM (2008) Drugs used to treat Parkinson's disease, present status and future directions. CNS Neurol Disord Drug Targets 7:321–342

Adewumi O, Aflatoonian B, Ahrlund-Richter L, Amit M, Andrews PW, Beighton G, Bello PA, Benvenisty N, Berry LS, Bevan S et al (2007) Characterization of human embryonic stem cell lines by the International Stem Cell Initiative. Nat Biotechnol 25:803–816

Akiyama K, Chen C, Wang D, Xu X, Qu C, Yamaza T, Cai T, Chen W, Sun L, Shi S (2012) Mesenchymal-stem-cell-induced immunoregulation involves FAS-ligand-/FAS-mediated T cell apoptosis. Cell Stem Cell 10:544–555

Alviano F, Fossati V, Marchionni C, Arpinati M, Bonsi L, Franchina M, Lanzoni G, Cantoni S, Cavallini C, Bianchi F, Tazzari PL, Pasquinelli G, Foroni L, Ventura C, Grossi A, Bagnara GP (2007) Term amniotic membrane is a high throughput source for multipotent mesenchymal stem cells with the ability to differentiate into endothelial cells in vitro. BMC Dev Biol 7:11

Anderson WS, Lenz FA (2006) Surgery insight: deep brain stimulation for movement disorders. Nat Clin Pract Neurol 2:310–320

Anderson WS, O'Hara S, Lawson HC, Treede RD, Lenz FA (2006) Plasticity of pain-related neuronal activity in the human thalamus. Prog Brain Res 157:353–364

Arnold SE, Hyman BT, Flory J, Damasio AR, Van Hoesen GW (1991) The topographical and neuroanatomical distribution of neurofibrillary tangles and neuritic plaques in the cerebral cortex of patients with Alzheimer's disease. Cereb Cortex 1:103–116

Avasthi S, Srivastava RN, Singh A, Srivastava M (2008) Stem cell: past, present and future—a review article. IJMU 3:22–30

Baizabal JM, Magaril MF, Jesu SO, Covarrubias L (2003) Neural stem cells in development and regenerative medicine. Arch Med Res 34:572–588

Bajada S, Mazakova I, Richardson JB, Ashammakhi N (2008) Updates on stem cells and their applications in regenerative medicine. J Tissue Eng Regen Med 2:169–183

Becker AJ, McCulloch EA, Till JE (1963) Cytological demonstration of the clonal nature of spleen colonies derived from transplanted mouse marrow cells. Nature 197:452–454

Behrstock S, Ebert A, McHugh J, Vosberg S, Moore J et al (2006) Human neural progenitors deliver glial cell line-derived neurotrophic factor to parkinsonian rodents and aged primates. Gene Ther 13:379–388

Berchtold NC, Cotman CW (1998) Evolution in the conceptualization of dementia and Alzheimer's disease: Greco-Roman period to the 1960s. Neurobiol Aging 19:173–189

Birks J (2006) Cholinesterase inhibitors for Alzheimer's disease. Cochrane Database Syst Rev 1:CD005593

Blundell R, Shah M (2015) Neurodegenerative diseases and stem cell transplantation. J Stem Cell Res Ther 5:277. https://doi.org/10.4172/2157-7633.1000277

Brehm M, Zeus T, Kostering M, Kogler G, Wernet P, Strauer BE (2002) Intracoronary transplantation of autologous bone-marrow cells for therapeutic angiogenesis in patients with myocardial infarction. Eur J Clin Investig 32(2):77

Brignier AC, Gewirtz AM (2010) Embryonic and adult stem cell therapy. J Allergy Clin Immunol 25:S336–S344

Busch SA, Hamilton JA, Horn KP, Cuascut FX, Cutrone R, Lehman N, Deans RJ, Ting AE, Mays RW, Silver J (2011) Multipotent adult progenitor cells prevent macrophage-mediated axonal dieback and promote regrowth after spinal cord injury. J Neurosci 31:944–953

Cacciatore I, Baldassarre L, Fornasari E, Mollica A, Pinnen F (2012) Recent advances in the treatment of neurodegenerative diseases based on GSH delivery systems. Oxidative Med Cell Longev 2012:1–12

Caplan AI, Dennis JE (2006) Mesenchymal stem cells as trophic mediators. J CellBiochem 98:1076–1084

Castillo-Melendez M, Yawno T, Jenkin G, Miller SL (2013) Stem cell therapy to protect and repair the developing brain: a review of mechanisms of action of cord blood and amnion epithelial derived cells. Front Neurosci 7:194

Chang YJ, Hwang SM, Tseng CP, Cheng FC, Huang SH, Hsu LF, Hsu LW, Tsai MS (2010) Isolation of mesenchymal stem cells with neurogenic potential from the mesoderm of the amniotic membrane. Cells Tissues Organs 192:93–105

Chapel A, Bertho JM, Bensiodhoum M, Fouillard L, Young RG, Frick J, Demarquay C, Cuvelier F, Mathieu E, Trompier F, Dudoignon N, Germain C, Mazurier C, Aigueperse J, Borneman J, Gorin NC, Goumelon P, Thierry D (2003) Mesenchymal stem cells home to injured tissues when co-infused with hematopoietic cells to treat a radiation-induced multi organ failure syndrome. J Gene Med 5:1028–1038

Chaudhuri KR, Schapira AH (2009) Non-motor symptoms of Parkinson's disease: dopaminergic pathophysiology and treatment. Lancet Neurol 8:464–474

Chavakis E, Urbich C, Dimmeler S (2008) Homing and engraftment of progenitor cells: a prerequisite for cell therapy. J Mol Cell Cardiol 45:514–522

Chen T, You Y, Jiang H, Wang ZZ (2017) Epithelial–mesenchymal transition (EMT): a biological process in the development, stem cell differentiation, and tumorigenesis. J Cell Physiol 232:3261–3272

Choi MR, Kim HY, Park JY, Lee TY, Baik CS, Chai YG, Jung KH, Park KS, Roh W, Kim KS, Kim SH (2010) Selection of optimal passage of bone marrow-derived mesenchymal stem cells for stem cell therapy in patients with amyotrophic lateral sclerosis. Neurosci Lett 472:94–98

Chuang J-H, Tung L-C, Lin Y (2015) Neural differentiation from embryonic stem cells in vitro: an overview of the signaling pathways. World J Stem Cells 7(2):437–447

Clarke DL, Johansson CB, Wilbertz J, Veress B, Nilsson E, Karlström H, Lendahl U, Frisén J (2000) Generalized potential of adult neural stem cells. Science 288:1660–1663

Cselenyak A, Pankotai E, Horvath EM, Kiss L, Lacza Z (2010) Mesenchymal stem cells rescue cardiomyoblasts from cell death in an in vitro ischemia model via direct cell-to-cell connections. BMC Cell Biol 11:29

da Silva Meirelles L, Caplan AI, Nardi NB (2008) In search of the in vivo identity of mesenchymal stem cells. Stem Cells 26:2287–2299

Daar AS, Greenwood HL (2007) A proposed definition of regenerative medicine. J Tissue Eng Regen Med 1:179–184

Das S, Zhou K, Ghosh D et al (2016) Implantable amyloid hydrogels for promoting stem cell differentiation to neurons. NPG Asia Mater 8:e304. https://doi.org/10.1038/am.2016.116

De Coppi P, Bartsch GJ, Siddiqui MM, Xu T, Santos CC, Perin L, Mostoslavsky G, Serre AC, Snyder EY, Yoo JJ, Furth ME, Soker S, Atala A (2007) Isolation of amniotic stem cell lines with potential for therapy. Nat Biotechnol 25:100–106

de Girolamo L, Lucarelli E, Alessandri G, Avanzini MA, Bernardo ME, Biagi E, Brini AT, D'Amico G, Fagioli F, Ferrero I et al (2013) Mesenchymal stem/stromal cells: a new "cells as drugs" paradigm. Efficacy and critical aspects in cell therapy. Curr Pharm Des 19:2459–2473

Deng J, Petersen BE, Steindler DA, Jorgensen ML, Laywell ED (2006) Mesenchymal stem cells spontaneously express neural proteins in culture and are neurogenic after transplantation. Stem Cells 24:1054–1064

Dharmasaroja P (2009) Bone marrow-derived mesenchymal stem cells for the treatment of ischemic stroke. J Clin Neurosci 16:12–20

El-Hashash AH (2016a) Stem cells, developmental biology and reparative/regenerative medicine: tools and hope for better human life. In: El-Hashash A (ed) Developmental and stem cell biology in health and disease. Bentham Science Publisher, Madison, pp 3–5
El-Hashash AH (2016b) Neural crest stem cells: a hope for neural regeneration. In: Abdelalim E (ed) Recent advances in stem cells: from basic research to clinical applications. Springer, Berlin, pp 233–250
Evans MJ, Kaufman MH (1981) Establishment in culture of pluripotential cells from mouse embryos. Nature 292:154–156
Farooqui AA, Horrocks LA (2007) Glycerophospholipids in the brain: phospholipases A2 in neurological disorders. Springer, New York
Fasano M, Alberio T, Lopiano L (2008) Peripheral biomarkers of Parkinson's disease as early reporters of central neurodegeneration. Biomark Med 2:465–478
Fearnley JM, Lees AJ (1991) Ageing and Parkinson's disease: substantia nigra regional selectivity. Brain 114:2283–2301
Fibbe WE, Nauta AJ, Roelofs H (2007) Modulation of immune responses by mesenchymal stem cells. Ann N Y Acad Sci 1106:272–278
Forte E, Chimenti I, Rosa P (2017) EMT/MET at the crossroad of stemness, regeneration and oncogenesis: the Ying-Yang equilibrium recapitulated in cell spheroids. Cancers 9(8):98. https://doi.org/10.3390/cancers9080098
Frenck R, Blackburn E, Shannon K (1998) The rate of telomere sequence loss in human leukocytes varies with age. Proc Natl Acad Sci U S A 95:5607–5610
Gai WP, Blessing WW, Blumbergs PC (1995) Ubiquitin-positive degenerating neurites in the brainstem in Parkinson's disease. Brain 118:1447–1459
Galindo LT, Filippo TR, Semedo P, Ariza CB, Moreira CM, Camara NO, Porcionatto MA (2011) Mesenchymal stem cell therapy modulates the inflammatory response in experimental traumatic brain injury. Neurol Res Int 2011, 564089 9 pages
Galli R, Borello U, Gritti A, Minasi MG, Bjornson C, Coletta M et al (2000) Skeletal myogenic potential of human and mouse neural stem cells. Nat Neurosci 3:986–991
Gardner RL (2007) Stem cells and regenerative medicine: principles, prospects and problems. C R Biol 330:465–473
Gerlach M, Maetzler W, Broich K, Hampel H, Rems L, Reum T, Riederer P, Stöffler A, Streffer J, Berg D (2012) Biomarker candidates of neurodegeneration in Parkinson's disease for the evaluation of disease-modifying therapeutics. J Neural Transm 119:39–52
Gomez-Isla T, Hollister R, West H, Mui S, Growdon JH, Petersen RC, Parisi JE, Hyman BT (1997) Neuronal loss correlates with but exceeds neurofibrillary tangles in Alzheimer's disease. Ann Neurol 41:17–24
Grade S, Götz M (2017) Neuronal replacement therapy: previous achievements and challenges ahead. NPJ Regen Med 2(29):1–11
Han F, Baremberg D, Gao J et al (2015) Development of stem cell-based therapy for Parkinson's disease. Transl Neurodegen 4:16
Hatzimichael E, Tuthill M (2010) Hematopoietic stem cell transplantation. Stem Cells Cloning 3:105–117
He S, Chen J, Zhang Y et al (2017) Sequential EMT-MET induces neuronal conversion through Sox2. Cell Discov 3:17017. https://doi.org/10.1038/celldisc.2017.17
Himes BT, Neuhuber B, Coleman C, Kushner R, Swanger SA, Kopen GC, Wagner J, Shumsky JS, Fischer I (2006) Recovery of function following grafting of human bone marrow-derived stromal cells into the injured spinal cord. Neurorehabil Neural Repair 20:278–296
Ho HY, Li M (2006) Potential application of embryonic stem cells in Parkinson's disease: drug screening and cell therapy. Regen Med 1:175–182
Hu X et al (2016) Identification of a common mesenchymal stromal progenitor for the adult haematopoietic niche. Nat Commun 7:13095. https://doi.org/10.1038/ncomms13095
Jackson KA, Mi T, Goodell MA (1999) Hematopoietic potential of stem cells isolated from murine skeletal muscle. Proc Natl Acad Sci U S A 96:14482–14486

Jankovic J, Aguilar LG (2008) Current approaches to the treatment of Parkinson's disease. Neuropsychiatr Dis Treat 4:743–757. https://doi.org/10.2147/NDT.S2006
Jellinger KA (2010) Basic mechanisms of neurodegeneration: a critical update. J Cell Mol Med 14:457–487
Jones R, Lebkowski J, McNiece I (2010) Stem cells. Biol Blood Marrow Transplant 16:S115–S118
Kim D, Chun BG, Kim YK, Lee YH, Park CS, Jeon I, Cheong C, Hwang TS, Chung H, Gwag BJ, Hong KS, Song J (2008) In vivo tracking of human mesenchymal stem cells in experimental stroke. Cell Transplant 16:1007–1012
Kim HJ, Lee JH, Kim SH (2010) Therapeutic effects of human mesenchymal stem cells on traumatic brain injury in rats: secretion of neurotrophic factors and inhibition of apoptosis. J Neurotrauma 27:131–138
Kim YS, Kim YK, Hwang O, Kim DJ (2012) Pathology of neurodegenerative diseases. In: Gonzalez-Quevedo A (ed) Brain damage—bridging between basic research and clinics. InTech, Croatia, pp 99–138
Kish SJ, Shannak K, Hornykiewicz O (1988) Uneven pattern of dopamine loss in the striatum of patients with idiopathic Parkinson's disease: pathophysiologic and clinical implications. N Engl J Med 318:876–880
Kofidis T, deBruin JL, Yamane T, Balsam LB et al (2004) IGF-1 promotes engraftment, differentiation and functional improvement after transfer of embryonic stem cells for myocardial restoration. Stem Cells 22:1239–1245
Kostic V, Przedborski S, Flaster E, Sternic N (1991) Early development of levodopa-induced dyskinesias and response fluctuations in young-onset Parkinson's disease. Neurology 41:202–205
Lagasse E, Connors H, Al-Dhalimy M, Reitsma M, Dohse M, Osborne L, Wang X, Finegold M, Weissman IL, Grompe M (2000) Purified hematopoietic stem cells can differentiate into hepatocytes in vivo. Nat Med 6:1229–1234
Lanctôt K, Rajaram RD, Herrmann N (2009) Review: therapy for Alzheimer's disease: how effective are current treatments? Ther Adv Neurol Disord 2:163–180
Landry DW, Zucker HA (2004) Embryonic death and the creation ofhuman embryonic stem cells. J Clin Invest 114:1184–1186
Le Blanc K (2003) Immunomodulatory effects of fetal and adult mesenchymal stem cells. Cytotherapy 5:485–489
Le Blanc K, Ringden O (2007) Immunomodulation by mesenchymal stem cells and clinical experience. J Intern Med 262:509–525
Lee EH, Hui JH (2006) The potential of stem cells in orthopaedic surgery. J Bone Jt Surg 88:841–853
Lee PH, Park HJ (2009) Bone marrow-derived mesenchymal stem cell therapy as a candidate disease-modifying strategy in Parkinson's disease and multiple system atrophy. J Clin Neurol 5:1–10
Levi B, James AW, Nelson ER, Vistnes D, Wu B, Lee M, Gupta A, Longaker MT (2010) Human adipose derived stromal cells heal critical size mouse calvarial defects. PLoS One. 5:11177
Li L, Xia Y (2014) Study of adipose tissue-derived mesenchymal stem cells transplantation for rats with dilated cardiomyopathy. Ann Thorac Cardiovasc Surg 20(5):398–406
Li SF, Lu XF, Sun MH (2002) Biological characteristics of mesenchymal stem cells in vitro derived from bone marrow of banna minipig inbred line. Zhongguo Xiu Fu Chong Jian Wai Ke Za Zhi 16:354–358
Li P, Yang B, Rand Gao W-Q (2014) Contributions of epithelial-mesenchymal transition and cancer stem cells to the development of castration resistance of prostate cancer. Mol Cancer 13:55
Liechty KW, MacKenzie TC, Shaaban AF, Radu A et al (2000) Human mesenchymal stem cells engraft and demonstrate site-specific differentiation after in utero transplantation in sheep. Nat Med 6:1282–1286
Lindvall O, Kokaia Z, Martinez-Serrano A (2004) Stem cell therapy for human neurodegenerative disorders.how to make it work. Nat Med 10:S42–S50
Liu YH, Ravi K, Wu SM (2008) Cardiovascular stem cells in regenerative medicine: ready for prime time? Drug Discov Today Ther Strateg 5:201

Lu D, Mahmood A, Wang L, Li Y, Lu M, Chopp M (2001) Adult human bone marrow stromal cells administered intravenously to rats after traumatic brain injury migrate into brain and improve neurological outcome. Neuroreport 12:559–563
Ma M, Sha C, Zhou Z, Zhou Q, Li Q (2008) Generation of patient-specific pluripotent stem cells and directed differentiation of embryonic stem cells for regenerative medicine. J Nanjing Med Univ 22:135–142
Mahmood A, Lu D, Chopp M (2004) Intravenous administration of marrow stromal cells (MSCs) increases the expression of growth factors in rat brain after traumatic brain injury. J Neurotrauma 21:33–39
Markov V, Kusumi K, Tadesse MG, William DA, Hall DM, Lounev V, Carlton A, Leonard J, Cohen RI, Rappaport EF, Saitta B (2007) Identification of cord blood-derived mesenchymal stem/stromal cell populations with distinct growth kinetics, differentiation potentials, and gene expression profiles. Stem Cells Dev 16:53–73
Martin GR (1981) Isolation of a pluripotent cell line from early mouse embryos cultured in medium conditioned by teratocarcinoma stem cells. Proc Natl Acad Sci U S A 78:7634–7638
Martin MJ, Muotri A, Gage F, Varki A (2005) Human embryonic stem cells express an immunogenic nonhuman sialic acid. Nat Med 11:228–232
Mayeux R (2003) Epidemiology of neurodegeneration. Annu Rev Neurosci 26:81–104
Mezey E, Chandross KJ, Harta G, Maki RA, McKercher SR (2000) Turning blood into brain: cells bearing neuronal antigens generated in vivo from bone marrow. Science 290:1779–1782
Miki T, Lehmann T, Cai H, Stolz DB, Strom SC (2006) Stem cell characteristics of amniotic epithelial cells. Stem Cells 23:1549–1559
Mizuno H (2009) Adipose-derived stem cells for tissue repair and regeneration: ten years of research and a literature review. J Nippon Med Sch 76:56–66
Munoz JL, Greco SJ, Patel SA, Sherman LS, Bhatt S, Bhatt RS, Shrensel JA, Guan YZ, Xie G, Ye JH, Rameshwar P, Siegel A (2012) Feline bone marrow-derived mesenchymal stromal cells (MSCs) show similar phenotype and functions with regards to neuronal differentiation as human MSCs. Differentiation 84:214–222
Nagal A, Singla RK (2012) Parkinson's disease: diagnosis, therapeutics and management. WebmedCentral Pharmaceutic Sci 3:WMC00 3670
Nauta AJ, Fibbe WE (2007) Immunomodulatory properties of mesenchymal stromal cells. Blood 110:3499–3506
Oh JY, Kim MK, Shin MS, Lee HJ, Ko JH, Wee WR, Lee JH (2008) The anti-inflammatory and antiangiogenic role of mesenchymal stem cells in corneal wound healing following chemical injury. Stem Cells 26:1047–1055
Olanow CW, Perl DP, DeMartino GN, McNaught KS (2004) Lewy-body formation is an aggresome-related process: a hypothesis. Lancet Neurol 3:496–503
Park HJ, Lee PH, Bang OY, Lee G, Ahn YH (2008) Mesenchymal stem cells therapy exerts neuroprotection in a progressive animal model of Parkinson's disease. J Neurochem 107:141
Parkinson J (2002) An essay on the shaking palsy. 1817. J Neuropsychiatry Clin Neurosci 14:223–236
Parr AM, Tator CH, Keating A (2007) Bone marrow-derived mesenchymal stromal cells for the repair of central nervous system injury. Bone Marrow Transplant 40:609–619
Piccini P, Pavese N, Hagell P, Reimer J, Bjorklund A et al (2005) Factors affecting the clinical outcome after neural transplantation in Parkinson disease. Brain 128:2977–2986
Polak J, Mantalaris S, Harding SE (2008) Advances in tissue engineering. Imperial College Press, London, pp 1–903
Ponte AL, Marais E, Gallay N, Langonné A, Delorme B, Hérault O, Charbord P, Domenech J (2007) The in vitro migration capacity of human bone marrow mesenchymal stem cells: comparison of chemokine and growth factor chemotactic activities. Stem Cells 25:1737–1745
Potapova IA, Gaudette GR, Brink PR, Robinson RB, Rosen MR, Cohen IS, Doronin SV (2007) Mesenchymal stem cells support migration, extracellularmatrix invasion, proliferation, and survival of endothelial cells in vitro. Stem Cells 25:1761–1768

Prince M, Bryce R, Albanese E, Wimo A, Ribeiro W, Ferri CP (2013) The global prevalence of dementia: a systematic review and metaanalysis. Alzheimers Dement 9:63–75

Qian D, Wei G, Xu C, He Z et al (2016) Bone marrow-derived mesenchymal stem cells (BMSCs) repair acute necrotized pancreatitis by secreting microRNA-9 to target the NF-κB1/p50 gene in rats. Nat Sci Rep 7:581. https://doi.org/10.1038/s41598-017-00629-3

Querfurth HW, LaFerla FM (2010) Alzheimer's Disease. N Engl J Med 362:329–344

Ramakrishna V, Janardhan PB, Sudarsanareddy L (2011) Stem cells and regenerative medicine—a review. Ann Rev Res Biol 1:79–110

Reyes M, Lund T, Lenvik T, Aguiar D, Koodie L, Verfaillie CM (2001) Purification and ex vivo expansion of postnatal human marrow mesodermal progenitor cells. Blood 98:2615–2625

Rigamonti A, Repetti GG, Sun C et al (2016) Large-scale production of mature neurons from human pluripotent stem cells in a three-dimensional suspension culture system. Stem Cell Rep 6(6):993–1008

Rocca WA, Petersen RC, Knopman DS, Hebert LE, Evans DA, Hall KS, Gao S, Unverzagt FW, Langa KM, Larson EB, White LR (2011) Trends in the incidence and prevalence of Alzheimer's disease, dementia, and cognitive impairment in the United States. Alzheimers Dement 7:80–93

Rubinsztein DC (2006) The roles of intracellular protein-degradation pathways in neurodegeneration. Nature 443:780–786

Rufer N, Brummendorf TH, Kolvraa S et al (1999) Telomere fluorescence measurements in granulocytes and T lymphocyte subsets point to a high turnover of hematopoietic stem cells and memory T cells in early childhood. J Exp Med 190:157–167

Sanberg PR (2007) Neural stem cells for Parkinson's disease: to protect and repair. Proc Natl Acad Sci U S A 104:11869–11870

Sanchez-Ramos J, Song S, Cardozo-Pelaez F, Hazzi C, Stedeford T, Willing A, Freeman TB, Saporta S, Janssen W, Patel N, Cooper DR, Sanberg PR (2000) Adult bone marrow stromal cells differentiate into neural cells in vitro. Exp Neurol 164:247–256

Seo JH, Cho SR (2012) Neurorestoration induced by mesenchymal stem cells: potential therapeutic mechanisms for clinical trials. Yonsei Med J 53:1059–1067

Shamblott MJ, Axelman J, Wang S, Bugg EM, Littlefield JW, Donovan PJ, Blumenthal PD, Huggins GR, Gearhart JD (1998) Derivation of pluripotent stem cells from cultured human primordial germ cells. PNAS 95:13726–13731

Sheyn D, Mizrahi O, Benjamin S, Gazit Z, Palled G, Gazit D (2010) Genetically modified cells in regenerative medicine and tissue engineering. Adv Drug Deliv Rev 62(7–8):683–698. https://doi.org/10.1016/j.addr.2010.01.002

Shin JY, Park HJ, Kim HN et al (2014) Mesenchymal stem cells enhance autophagy and increase β-amyloid clearance in Alzheimer disease models. Autophagy 10(1):32–44. https://doi.org/10.4161/auto.26508

Si YL, Zhao YL, Hao HJ, Fu XB, Han WD (2011) MSCs: biological characteristics, clinical applications and their outstanding concerns. Ageing Res Rev 10:93–103

Singh P, Williams DJ (2008) Cell therapies: realizing the potential of this new dimension to medical therapeutics. J Tissue Eng Regen Med 2:307–319

Speth JM, Hoggatt J, Singh P, Pelus LM (2014) Pharmacologic increase in HIF1α enhances hematopoietic stem and progenitor homing and engraftment. Blood 123(2):203–207

Steiner AF, Karl N, Pilapil C, Noth U, Evans CH, Murray MM (2006) Multilineage mesenchymal differentiation potential of cells migrating out of the anterior cruciate ligament. In: Transactions of the 52nd annual meeting of the Orthopaedic Research Society, Chicago, IL. p 1133

Tabert MH, Liu X, Doty RL, Serby M, Zamora D, Pelton GH, Marder K, Albers M, Stern Y, Devanand DP (2005) A 10-item smell identification scale related to risk for Alzheimer's disease. Ann Neurol 58:155–160

Takagi Y, Takahashi J, Saiki H et al (2005) Dopaminergic neurons generated from monkey embryonic stem cells function in a Parkinson primate model. J Clin Invest 115:102–109

Tayeb HO, Yang HD, Price BH, Tarazi FI (2012) Pharmacotherapies for Alzheimer's disease: beyond cholinesterase inhibitors. Pharmacol Ther 134:8–25

Terrigno M, Busti I, Alia C, Pietrasanta M, Arisi I, D'Onofrio M, Caleo M, Cremisi F (2018) Neurons generated by mouse ESCs with hippocampal or cortical identity display distinct projection patterns when co-transplanted in the adult brain. Stem Cell Rep 10(3):1016–1029
Thiery JP, Acloque H, Ruby YJ, Huang M, Nieto A (2009) Epithelial-mesenchymal transitions in development and disease. Cell 139(5):871–890
Thomson JA, Eldor JI, Shapiro SS, Waknitz MA, Swiergiel JJ, Marshall VS, Jones JM (1998) Embryonic stem cell lines derived from human blastocysts. Science 282:1145–1147
Togel F, Weiss K, Yang Y, Hu Z, Zhang P, Westenfelder C (2007) Vasculotropic, paracrine actions of infused mesenchymal stem cells are important to the recovery from acute kidney injury. Am J Physiol Renal Physiol 292:F1626–F1635
Ulrike M, Konrad M (2003) Alzheimer: the life of a physician and the career of a disease. Columbia University Press, New York, p 270
van Velthoven CT, Kavelaars A, van Bel F, Heijnen CJ (2010a) Repeated mesenchymal stem cell treatment after neonatal hypoxia-ischemia has distinct effects on formation and maturation of new neurons and oligodendrocytes leading to restoration of damage, corticospinal motor tract activity, and sensorimotor function. J Neurosci 30(28):9603–9611
van Velthoven CT, Kavelaars A, van Bel F, Heijnen CJ (2010b) Mesenchymal stem cell treatment after neonatal hypoxic-ischemic brain injury improves behavioral outcome and induces neuronal and oligodendrocyte regeneration. Brain Behav Immun 24(3):387–393
Waldemar G, Dubois B, Emre M, Georges J, McKeith IG, Rossor M, Scheltens P, Tariska P, Winblad B (2007) Recommendations for the diagnosis and management of Alzheimer's disease and other disorders associated with dementia: EFNS guideline. Eur J Neurol 14:e1–e26
Weissman IL (2000) Stem cells: units of development, units of regeneration, and units in evolution. Cell 100:157–168
Woodbury D, Schwarz EJ, Prockop DJ, Black IB (2000) Adult rat and human bone marrow stromal cells differentiate into neurons. J Neurosci Res 61:364–370
Wu C-C et al (2016) Gain of BDNF function in engrafted neural stem cells promotes the therapeutic potential for Alzheimer's disease. Sci Rep 6:27358. https://doi.org/10.1038/srep27358
Xin H, Li Y, Shen LH, Liu X, Wang X, Zhang J, Pourabdollah-Nejad DS, Zhang C, Zhang L, Jiang H, Zhang ZG, Chopp M (2010) Increasing tPA activity in astrocytes induced by multipotent mesenchymal stromal cells facilitate neurite outgrowth after stroke in the mouse. PLoS One 5:9027
Ye M, Wang XJ, Zhang YH, Lu GQ, Liang L, Xu JY, Chen SD (2007) Therapeutic effects of differentiated bone marrow stromal cell transplantation on rat models of Parkinson's disease. Parkinsonism Relat Disord 13:44–49
Yi ES, Lee SH, Son MH, Kim JY, Cho EJ, Lim SJ, Cheuh HW, Yoo KH, Sung KW, Koo HH (2012) Hematopoietic stem cell transplantation in children with acute leukemia: similar outcomes in recipients of umbilical cord blood versus marrow or peripheral blood stem cells from related or unrelated donors. Korean J Pediatr 55:93–99
Yue C, Jing N (2015) The promise of stem cells in the therapy of Alzheimer's disease. Transl Neurodegener 4:8
Zhang J, Li Y, Chen J, Yang M, Katakowski M, Lu M, Chopp M (2004) Expression of insulin-like growth factor 1 and receptor in ischemic rats treated with human marrow stromal cells. Brain Res 1030:19–27
Zhang J, Li Y, Lu M, Cui Y, Chen J, Noffsinger L, Elias SB, Chopp M (2006) Bone marrow stromal cells reduce axonal loss in experimental autoimmune encephalomyelitis mice. J Neurosci Res 84:587–595

Chapter 10
Stem Cells for Nerve and Muscle Repair: Harnessing Developmental Dynamics in Therapeutics

Satish Sasikumar, Ashima Bhan, and T. K. Rajendra

Abbreviations

ALS	Amyotrophic lateral sclerosis
BBB	Blood brain barrier
CD	Cluster of differentiation
CNS	Central nervous system
CRISPR	Clustered regularly interspaced short palindromic repeats
DMD	Duchenne muscular dystrophy
DNA	Deoxyribonucleic acid
ENS	Enteric nervous system
ES cells	Embryonic stem cells
EVs	Extracellular vesicles
HRT	Hormone replacement therapies
IA	Intra-arterial
iPSCs	Induced pluripotent stem cell(s)
IV	Intra-venous
LSDs	Lysosomal storage disorders
MABs	Mesoangioblasts
MCAO	Middle cerebral artery occlusion
MDSCs	Muscle-derived stem cells
MEFs	Mouse embryonic fibroblasts
miRNAs	microRNAs
MPCs	Muscle progenitor cells
MSCs	Mesenchymal stem cells

The original version of this chapter was revised. The correction to this chapter is available at https://doi.org/10.1007/978-3-319-98065-2_11

S. Sasikumar (✉) · A. Bhan · T. K. Rajendra (✉)
Dr. D. Y. Patil Biotechnology and Bioinformatics Institute, Dr. D. Y. Patil Vidyapeeth, Pune, Maharashtra, India

P. V. Pham, A. El-Hashash (eds.), *Stem Cells for Cancer and Genetic Disease Treatment*, Stem Cells in Clinical Applications, https://doi.org/10.1007/978-3-319-98065-2_10

Myf5	Myogenic factor 5
MyoD	Myoblast determination protein
NMDs	Neuromuscular diseases/disorders
NPCs	Neural progenitor cells
NSCs	Neural stem cells
NSPCs	Neural stem/progenitor cells
PICs	$PW1^+$ interstitial cells
RICE	Rest, ice/cold, compression and elevation
RNA	Ribonucleic acid
RNAi	RNA interference
SMA	Spinal muscular atrophy
Sox	Sry-related High Mobility Group (HMG) box
SP	Side population
Sry	Sex determining region Y (present on the Y chromosome)
SVZ	Subventricular zone
3D	Three-dimensional

10.1 Introduction

Existence in the animal kingdom has meaning only in the context of the intellectual ability to think and respond to situations so as to live, procreate, and evolve. In any collective cellular conglomerate that constitutes the animal kingdom, critical activities like 'thinking' and effectively responding to situations are the primary functions of the nervous system. However, a thinking and conducting nervous system is a futile unit without the orchestrated functioning of the musculature for mounting an effective and observable response. Hence, coordination of neuromuscular functions is a must for the existence and evolution of life as, being mobile (excluding sedentary animals), animals have to not only fight for food and mating, but also have to go in search of or hunt for food with stiff competition. Animal life, in general, and evolution, in particular, would not have been possible without a well-oiled neuromuscular coordination, which have in fact provided great survival and fitness values. Overall, it can be argued that the animal life as we know could not have evolved into a very sophisticated intellectual human entity without putting in place a well-orchestrated neuromuscular system. However, a compromise in the functioning of the nervous and neuromuscular systems due to either infectious diseases, inherited disorders or lifestyle will limit those evolutionarily acquired critical advantages.

Diseases/Disorders, in general, can be classified into two broad categories: those that affect only the healthy lifestyle and ones that affect not only healthy lifestyle, but also the very existence and evolution of life. Though the severity and the age of onset of any disease/disorder affecting pre- or post-reproductive age can be a determining factor in their evolutionary outcomes, those affecting neuromusculature fall into both the categories and hence the importance of diseases/disorders affecting them. NMDs represent a heterogeneous group of diseases/disorders with varying severity and behavioral outcomes.

10.2 Neuronal and Neuromuscular Disorders

Neurological disorders can be broadly classified as those affecting sensory perception, memory, intellectual ability, hormonally controlled bodily functions, standard behavioral patterns, and motor functions when they affect neuromuscular coordination, though many of the affected classes may be interrelated. Any of these neurological dysfunctions could either be due to errors in metabolism or death of neurons with behavioral responses depending on the cell types affected (Schappi et al. 2013; Poe et al. 1989; van der Knaap and Valk 1988). Though neuromuscular disorders could also occur due to either metabolic errors or neuronal or muscle death, they are characterized by progressive muscle weakness and degeneration with or without neuronal involvement (Vincent 2008; Hirsch 2007; Orbay and Mizuno 2013). Though the primary defect in neuromuscular disorders may start either in the motor neuron or the target muscle, lack of activity in any one of the synaptic partners negatively affects the other synaptic partner. Treating any of these neuronal or neuromuscular disorders involves either classical pharmaceutical interventions or modern stem cell therapeutics. Classical pharmaceutical interventions for these disorders require, in addition to the knowledge of pharmacological principles, an understanding of basic neuronal and neuromuscular biology and the underlying causative principles. Stem cell therapeutics, however, requires, in addition to the basic information needed for pharmaceutical interventions, a deep understanding of stem cells and how the nervous and neuromuscular systems are established during development and differentiation.

10.3 Development and Differentiation of the Nervous and Neuromuscular Systems

The development and differentiation of nervous and neuromuscular systems could be discussed at three interdependent levels and they are as follows:

- Segregation and subsequent differentiation of embryonic cells into nervous and neuromuscular systems
- Establishment of left-right symmetry and proportionality
- Maintenance of neuromuscular health

Early embryonic cells carry tremendous plasticity to elaborate a body pattern by progressive and differential activation and inactivation of genes in a coordinated manner. Zygote can be considered to have the full or total potency (or more specifically, "totipotency") to give rise to a fully differentiated multicellular organism. This totipotency is retained in each cell of the 8-cell embryo in mammals and each of these cells can generate a fully differentiated individual, if appropriate conditions are available (Chan et al. 2000; Pera et al. 2000). During the process of differentiation and organogenesis, some tissue-specific cells are set aside to self-renew without undergoing differentiation and serve as adult stem cells. These stem cells persist throughout life by continuously renewing themselves and differentiate to replenish the lost or damaged cells for maintaining homeostasis in the tissue (Tweedell 2017; Jopling et al. 2011; Slack 2008; Jones and Wagers 2008; Spradling et al. 2001).

Placement of the nervous system and neuromusculature in the correct spatiotemporal geometry are two cardinal events during development and differentiation. While neurogenesis and myogenesis are independent events, their overall development is a concerted and interdependent process, especially while laying down neuromuscular and interneural connectivity. A layer of complexity in neural development is obvious when one looks at the mechanisms and genes involved in guiding axons to find their synaptic partner (pathfinding using extracellular molecular guidance cues), which can be a neuron or a muscle (Hessinger et al. 2017; Kao et al. 2017; Dun and Parkinson 2017; Short et al. 2016; Joo et al. 2013; Philippidou and Dasen 2013; Di Bonito et al. 2013; Bentzinger et al. 2012; Narita and Rijli 2009; Dasen and Jessell 2009; Sieburth et al. 2005; Verity et al. 2003; Muntoni et al. 2002; Nieto 1996; Tessier-Lavigne and Goodman 1996; Appel et al. 1995; Chu-LaGraff et al. 1995; Meadow et al. 1994; Tsuchida et al. 1994; Keshishian and Chiba 1993; van der Knaap and Valk 1988). While cardiac muscle development follows its own unique pattern, skeletal muscle development primarily involves somite-derived muscle precursor cells, or myoblasts, which undergo specification and differentiation into muscle fibers. In the process, these myoblasts finally fuse to form syncytial primary and secondary myotubes which mature subsequently into muscle fibers (Roux and Zaffran 2016; Doherty et al. 2011; Tajbakhsh 2005; Taylor 2002; Buckingham 2001; Srivastava and Olson 2000; Chen and Fishman 2000; Rohwedel et al. 1994). In the adult, skeletal muscle stem cells are a population of cells called "satellite cells" that express unique marker genes like *Pax7*, *cdh15*, and *Myf5*, encoding proteins, Paired box-7, M-cadherin and Myogenic factor 5, respectively. Under conditions of injury or exercise, these quiescent cells proliferate and contribute their nuclei to preexisting muscle fibers to either repair or to maintain muscle homeostasis (Collins et al. 2005; Seale et al. 2000; Cornelison and Wold 1997). Though unique specification and differentiation programs are also applicable to development of the CNS, self-regenerating NPCs contribute to the formation of neurons and glia by exiting cell cycle and spatiotemporally activating genes required for differentiation into neural and glial cells (Anderson 1993, 1997; McConnell 1995; Hatten et al. 1991). One way by which such spatio-temporal control of gene expression is executed during neurogenesis is by sequential binding of Sox factors to a common set of neural genes, which is dictated by developmentally imposed stage-specific constraints (Bergsland et al. 2011). Taken together with a great body of work, genetic control of nervous system development has provided better insights into molecular and pathological mechanisms affecting the nervous system (Lee and Pfaff 2001; Belgacem et al. 2016; Philippidou and Dasen 2013; Eade et al. 2012; Bertrand et al. 2002).

Most of the information gathered so far on the genes responsible for the establishment of neuromuscular systems came from gene knockout studies done in *Caenorhabditis elegans, Drosophila*, zebrafish, and mice (Koon and Chan 2017; Edens et al. 2015; Plantie et al. 2015; Kuzdas-Wood et al. 2014; Babin et al. 2014; Stewart et al. 2014; McVey et al. 2012; Kabashi et al. 2011; O'Kane 2011; Lloyd and Taylor 2010; Marder 2002; Birchmeier and Brohman 2000; Arnold and Braun 2000; Sabourin and Rudnicki 2000; Perry and Rudnick 2000; Naya and Olson 1999; Yun and Wold 1996; Keshishian and Chiba 1993; Trent et al. 1983). Additional information regarding neuromuscular development also came from studies on regeneration as in the case of salamanders (Morrison et al. 2006). Though an understanding of genes involved in developmental decisions is critical for any therapeutic intervention in modern medicine, it is to be appreciated that with the evolution of complexity in multicellular organisms, there arose mechanisms to establish and maintain size, proportionality, and symmetry in relation to their functional interdependency. Any agent that disturbs this multicellular fabrication is expected to cause a disease or disorder and hence establishing and maintaining tissue/organ size, proportionality and their functional relatedness is of utmost importance.

10.4 Size and Proportionality in Neuronal and Neuromuscular Health

Life is simple with single-celled organisms. However, with the advent of multicellularity, establishing and maintaining placement, shape, and size of tissues and organs in relation to the overall bodily parameters and function is a must for proper functionality. Relative size and proportionality of any tissue or an organ in relation to other parts of the body are specific within a range. For example, number of neurons in the brain, in general, or their number in any specific part of the nervous system, in particular, is tailored to match the overall body size and behavioral need of any given species (Nelson 2009). Such species-specific cell number, tissue and organ size is established and maintained by the genetically ordained ability of tissues to sense cell density, which is sometimes referred to as "quorum sensing", and respond accordingly (Penzo-Mendez and Stanger 2015; Schwank and Basler 2010; Stanger 2008; Day and Lawrence 2000). For example, if one were to look at the ratio of granule cells and Purkinje cells in mice and humans, it is less than 200 to 1 in the former and 3000 to 1 in the latter! (Wetts and Herrup 1982; Lange 1975). While these examples relate to evolutionarily and developmentally determined numerical cell and tissue proportionality, a glaring example in a pathological situation is the poliovirus induced poliomyelitis and the associated post-polio syndrome affecting body proportionality or left-right symmetry in the neuromuscular phenotype (Huang and Shih 2015; Baj et al. 2015; Swanson and McGavern 2015; Koyuncu et al. 2013; Nathanson 2008; Jubelt and Drucker 1999; Jubelt 1992; Jubelt and Cashman 1987). Other extremes in this context are inherited genetic disorders like SMA, ALS, etc., which not only affect motor neurons, but also the skeletal muscles, either directly or indirectly due to primary loss of innervation (Patten et al. 2014; Kiernan et al. 2011; Prior 2010; Lefebvre et al. 1995; Brzustowicz et al. 1990) with the ultimate effect of losing cell and tissue volume amassed during development.

Some of the examples given above are a cursory indication of the importance of establishing and maintaining cell and tissue organization and their tight functional relationship in the overall organismal homeostasis. A disturbance in this homeostasis is the primary cause of any disease, congenital disorders, injury, or aging. However, it is to be noted that during development, cell number, tissue size, muscle volume, or organ proportion is established and elaborated in functional and structural relation to other body parameters by employing specific spatio-temporal molecular events, which involves extensive intercellular communications and differential gene expression. These events are executed by creating molecular gradients in the developmental landscape like the one explained with the help of a "French Flag Model" (Dahmann et al. 2011; Jaeger and Martinez-Arias 2009; Wolpert 1968). Molecular gradients, as evinced in the French Flag Model, lay down spatial concentration gradients of biomolecules, dubbed "morphogens" that act as signals for cells in their immediate neighborhood to choose specific fates. Cells in the immediate vicinity of the gradient respond according to specific concentration of the morphogen by activating a unique set of developmental genes, thus fixing cell fates to elaborate a specific tissue or an organ. While this elaborate process of differentiation to make specific morphological structures is referred to as morphogenesis, maintaining their newly attained developmental state or morphology with their associated functions is referred to as morphostasis or tissue homeostasis (Yu et al. 2015; Dahmann et al. 2011; Schwank and Basler 2010; Nelson 2009). One of the well-studied signaling pathways that is involved in organ size control and morphostasis is the Hippo signaling pathway, which in addition is also involved in regulating the unique properties of stem cells (Wang et al. 2017b; Yu et al. 2015; Manzanares and Rodriguez 2013; Liu et al. 2012; Ramos and Camargo 2012). Any injury, infection, nutritional deficiency, food habits, lifestyle, aging, or genetic anomaly that affects either neuronal or neuromuscular morphogenesis during development, and/or morphostasis subsequently, will lead to corresponding diseases or disorders.

Though neuronal or muscle number, size and their proportion are laid down during development, it is important to note, as explained earlier, that health and survival of neurons or muscles are also dependent on active interaction between the two synaptic partners, between two neurons, or between a motor neuron and the muscle. If any one of the synaptic partners is affected, other components in the neural or neuromuscular signaling network will also be affected, thereby manifesting into a pathological outcome. Any causative agent, acquired or inherited, affecting neural or neuromuscular functions can have symmetric (and/or systemic) or asymmetric effects on a bilaterally symmetrical system and hence the consequent behavioral output, the latter being exemplified by poliomyelitis. A genetic cause could lead to a late onset of disorder if the concerned genes are not involved in early developmental decisions. On the other hand, early developmental errors, involving genetic and/or non-genetic causes lead to a disproportionate elaboration of body plan, often as a result of failures in morphogenetic events, which are known to ensure basic proportionality in the body plan and "quorum sensing" (Schwank and Basler 2010). Establishment of cell number, tissue identity, tissue size, proportionality, and maintenance of structural and functional identity are in general studied under morphogenesis and morphostasis, whose compromised integrity is the root cause of various diseases and disorders, including those of the neuromusculature.

10.5 Morphostasis Recapitulates Morphogenesis

Morphostasis (tissue homeostasis) involves maintaining not only tissue memory to preserve its identity and function, but also repairing damaged tissues after infection or injury (and building extra muscle mass in the context of physical exercise). While maintenance of tissue identity and function is ensured simply by preserving differential gene expression profile established during morphogenesis in development, maintaining regenerative tissues like blood or skin and building muscle mass due to exercise or in response to infections requires the participation of stem cells.

Morphogenesis and morphostasis are species-specific phenomenon, which are genetically encoded to integrate overall body size with its functionality. Species with less number of neurons may have to command their neurons to work harder to meet efficient execution of function than in ones with more neurons for the same function. For example, loss of a single neuron responsible for egg laying in nematodes is enough to make them reproductively dead (Trent et al. 1983). In contrast, there occurred no observable behavioral defect when the number of human retinal ganglion cells, or Purkinje cells of mice, is reduced by more than 90% (Wetts and Herrup 1983; Quigley et al. 1982). On the contrary, behavioral response is not affected when a decrease in the number of neuronal boutons in *Drosophila* is restored by a compensatory increase in the bouton size in the neuromuscular synapse and the consequent increase in their surface area (Menon et al. 2013; Franco et al. 2004; Chang and Balice-Gordon 2000; Lnenicka and Keshishian 2000; Budnik 1996). Further, a decrease in the number of neurons to 20,000, from an average of 40,000, in the *Drosophila* optic medulla had no observable behavioral deficit, thus highlighting the inherent plasticity of the system to cope up with a significant decrease in neuronal number (Fischbach and Heisenberg 1981). Hence, it is likely that neural plasticity to cope up with the complex functional and behavioral needs evolved with the increase in evolutionary complexity and congruent increase in neural cell number. While the ability of optic medulla to function efficiently even with a significantly reduced number of neurons is one way of looking at plasticity, another way of looking at the neuronal plasticity is through the prism of regeneration. It is suggested that the adult mammalian neurons have lost the ability to regenerate axons due to the ambient inhibitory environment cast by, for example, the glial scar and the lesion caused due to spinal cord injury (Horner and Fred 2000). In spite of this imposed inability of neurons to regenerate axons, functional recuperation is indeed observed. Such a recuperation, that too in an inhibitory environment, has been explained as due to restructuring of remaining circuitry which is dependent on axonal sprouting of either damaged or intact nerve fibers (Horner and Fred 2000; Ramer et al. 2000; Bernstein and Mary 1971). Regeneration and recovery of axons seen in such situations were stimulated and guided by cumulative and combinatorial treatments involving neurotrophic factors, removal of inhibitory factors and physical exercise (Nishio et al. 2009; Hamid and Ray 2008; Smith et al. 2006; Engesser-Cesar et al. 2005; Simonen et al. 2003; Kim et al. 2001; Namiki et al. 2000; McTigue et al. 1998; Ye and Houle 1997; Grill et al. 1997). Overall, plasticity of this kind seen in multicellular animals have evolved by dedicating more cells than what is required (redundancy) or setting aside a population of stem cells for efficient functioning of the system, which buffer or cushion against the manifestation of diseases and disorders. For example, as detailed above, loss or compromised functionality of some cells in a tissue due to either damage or disease can be compensated by the remaining unaffected cells or by differentiating stem cells to replace damaged or lost cells. This ability or plasticity of neurons to repair damage either due to cumulative and combinatorial treatments or by increasing their number in relation to any given function during the course of evolution, provided a great leverage to face onslaughts such as injury, infections, aging, etc. Diseases and disorders manifest when such plasticity is stretched beyond a threshold when morphostasis is no longer possible.

10.6 Diverse Disorders of Nervous and Muscular Systems

In any neurological or neuromuscular disorder, the underlying molecular or biochemical defect manifests in the overall failed coordination between one or more neurons or among neurons and the muscle, and hence, results in a decline in their function. Such biochemical and cellular coordination will also have effects on motor functions either due to complete collapse of motor functions or due to incoherent left-right coordination and consequent motor function deficit invariably affecting lifestyle; this, in the worst case scenario, leads to a sedentary life which further deteriorates due to wasting of the neuro-musculature. Though sedentary confinement due to motor dysfunction can have fringe effects on social interactions and self-esteem which might culminate in behavioral disorders, they can also happen independently of motor function deficits. For example, epileptic episodes, cognitive disabilities, amnesia/dementia, schizophrenia, autism spectrum disorders, congenital mental disorders like Down syndrome and Fragile-X mental retardation, blindness, deafness, etc. manifest without involving primary motor function deficits. Such disorders involving nervous system, with or without motor output, can broadly be categorized based on the principal systems affected as detailed below:

1. Pituitary gland
2. Interneural network in the CNS
3. Sensory neuropathies
4. Neuromuscular system

The dysfunction of pituitary gland could either occur due to congenital anomalies, tumors, infections, vascular defects, or accidental damage leading to brain insult (Lara-Velazquez et al. 2017). As pituitary dysfunction affects the routine and the way of life, therapeutic measures involve HRT, often with unwanted side effects (Asa and Ezzat 2002; Regal et al. 2001). Disorders affecting inter-neural networks involve almost everything from the loss or impaired abilities to register, retain, integrate, and recall memory in a meaningful way to those that elicit epileptic seizures (Lax et al.

2017; Carpio et al. 2016; Li et al. 2016a, b; Zueter and Zaiter 2015; Swanson and McGavern 2015; Jerath and Shy 2015; Bartt 2012; Zett et al. 2012; Nicolson 2008; Garne et al. 2008; Younger 2004; Verity et al. 2003; Moller 2001; Woolf and Salte 2000). As reflected in its name, sensory neuropathies involve pathogenesis affecting sensory perceptions manifested due to inborn genetic defects, infection, injury, or even due to secondary causes such as diabetes (Zhou et al. 2016; Jerath and Shy 2015; Dyck et al. 1993). Those disorders that affect neuromuscular system can have defects in either or both of the synaptic partners due to any of the various reasons detailed earlier, culminating in motor function failure exemplified by SMA, ALS, etc. (Engel et al. 2015; Renton et al. 2014; Al-Chalabi and Hardiman 2013; Laing 2012; Bertini et al. 2011; Andersen and Al-Chalabi 2011; Erichsen et al. 2009; Chau and Griffith 2005; Feldkotter et al. 2002; Lefebvre et al. 1995, 1997).

Just like any other disease or disorder, those affecting nervous or neuromusculature can also fall into broad categories (Carpio et al. 2016; Li et al. 2016a, b; Engel et al. 2015; Swanson and McGavern 2015; Zueter and Zaiter 2015; Al Bekairy et al. 2014; Jerath and Shy 2015; Cruza et al. 2014; Schappi et al. 2013; Laing 2012; Lorenzoni et al. 2012; Bartt 2012; Bertini et al. 2011; Noyola et al. 2010; Crum-Cianflone 2008; Nicolson 2008; Chau and Griffith 2005; Younger 2004; Garne et al. 2008; Verity et al. 2003; Poe et al. 1989; van der Knaap and Valk 1988; Johnson 1984). Medications, electric shock therapies, surgeries, cognitive and physical exercises, and physiotherapy are the major means by which neuronal and neuromuscular disorders are treated in general, irrespective of whether these disorders are caused due to injury, infection, nutritional deficiency, food habits, lifestyle, aging, or genetic anomaly (Loy et al. 2017; Finsterer and Zarrouk-Mahjoub 2016; Lossing et al. 2017; Sharifi 2013; Lyons 2011; Mian et al. 2010; Platt and Lachmann 2009; Schwalb and Hamani 2008; Enns and Huhn 2008; Kley et al. 2008; Pierret et al. 2008; Weisfeldt and Zieman 2007; Malone and Pandya 2006; Mayberg et al. 2005; Byrne 2005; Verity et al. 2003; Kokontis and Gutmann 2000; Bartholow 1982). While medications have various kinds of side effects on other bodily functions, neither medication nor other treatment methods fully cure the disease (Ibi et al. 2017; Moavero et al. 2017; Cockburn et al. 2017; Finsterer and Scorza 2017; Hryhorowicz et al. 2018; Solmi et al. 2017; Hilt et al. 2014; Marano et al. 2011; Muench and Hamer 2010; Tueth 1994; Segraves 1988). At best, these treatment methods alleviate suffering by treating symptoms, without eliminating the cause, except in cases involving vaccine treatment. Even when palliative therapeutics is available, they work only when target neurons and neuromuscular connections are still intact, as these cells and their connections are prone to get deteriorated when they are non-functional. In contrast, surgical treatments may lead to complete cure or, on the contrary, can potentially incapacitate the patient. None of the treatment methods, leaving aside vaccination, guarantees complete cure thus necessitating novel methods of medical interventions like CRISPR, a gene editing technology to correct an inherited genetic anomaly or stem cell engineering, a method to replace affected cells (Terns and Terns 2014; Mali et al. 2013).

10.7 Stem Cells

Stem cells and their biology are huge areas in their own right and they are discussed here with special emphasis on the disorders of the nerve and muscle. Neural stem cells are multipotent in nature with the ability to differentiate into neurons, astrocytes, and oligodendrocytes. Given their self-renewal property, just like any other type of stem cells, they can be expanded in culture. In contrast to neural stem cells, neural progenitor cells have limited capacity for self-renewal and these two classes of cells are collectively referred to as NSPCs. Apart from ES cells and iPSCs as the source, the main source of NSPCs are the SVZ of the lateral ventricles, hippocampal dentate gyrus of the brain, and the ependymal lining of spinal cord central canal (Hu et al. 2010; Tropepe et al. 2001; Palmer et al. 1997; Weiss et al. 1996; Morshead et al. 1994).

Adult skeletal muscle is composed primarily of syncytial/multinucleate myofibers, each enveloped by a basal lamina. As described earlier, a population of muscle stem cells, called satellite cells, is found between the muscle fiber plasma membrane and the basal lamina (Wang et al. 2014; Shi and Garry 2006; Mauro 1961). Satellite cells are quiescent and can be identified by the expression of Pax7 and Integrin alpha-7 (Seale et al. 2000; Burkin and Kaufman 1999). Upon insult, these satellite cells get activated with the expression of myogenic regulatory factors like Myf5 and MyoD and re-enter the cell cycle, differentiate, and ultimately fuse with the damaged fibers to repair (Montarras et al. 2005; Relaix et al. 2005; Mauro 1961). Though satellite cells are the major source of skeletal muscle stem cells, other cell populations with myogenic capacity are also known to exist (see subsection 10.14 for details on transdifferentiation).

10.8 Plasticity, Morphostasis, and Stem Cell Therapeutics

Neuronal plasticity with the developmentally derived number of neurons is the major way by which the nervous system can withstand various disease causing conditions. This is because, in contrast to neurons, tissue damage in most other tissues can be re-established by reprogramming the remaining cells of the tissue or by employing stem cells. On the contrary, being post-mitotic and terminally differentiated, neither are the neurons able to divide nor are there enough precursor cells/stem cells that can replace lost neurons, in general (Miale and Sidman 1961; Rakic 1985). Further, wherever stem cells are available, they contribute towards rebuilding neural architecture through their "neural stem cell plasticity," which is different from "neural plasticity" as explained earlier (Zelentsova et al. 2017; Ottoboni et al. 2017; Baser et al. 2017). Thus, neuronal plasticity and morphostasis, in addition to neural and muscle "stem cell plasticity," are the cardinal principles affected in neuronal and neuromuscular disorders, be it due to infection, aging, or any other causative principle (Ottoboni et al. 2017; Brack and Munoz-Canoves 2016; Thorley et al. 2015; Breunig et al. 2015; Sierra et al. 2015; Blau et al. 2015; Lu et al. 2014; Ruller et al. 2012; Chakkalakal et al. 2012; Liu and Rando 2011; Shi and Garry 2006). Taken collectively, terminally differentiated cells of the nervous and neuromuscular systems have limited flexibility in repairing damage beyond what is allowed by their inherent plasticity.

The plasticity conferred upon nervous and neuromuscular systems by respective stem cells is due to their ability to invoke differential gene regulation and developmental decisions of early development specific to these organ systems. That means, in order to retain tissue plasticity and homeostasis, the endogenous stem cells have to have immaculate genetic programs to get modified into structurally and functionally relevant tissues, essentially recapitulating developmental events. In other words, and to be successful in stem cell therapeutics, one should be able to impose these tissue-specific developmental events to elaborate a cell or a tissue type of an optimal developmental stage by activating appropriate signaling molecules at proper locations (Chal et al. 2015; Blau et al. 2001, 2015; Bentzinger et al. 2012; Doherty et al. 2011; Sieburth et al. 2005; Chien 2004; Verity et al. 2003; Eriksson et al. 1998; Wise et al. 1996; Nieto 1996; Tessier-Lavigne and Goodman 1996; Appel et al. 1995; Chu-LaGraff et al. 1995; Rohwedel et al. 1994; Meadow et al. 1994; Tsuchida et al. 1994; Rohwedel et al. 1994; Keshishian and Chiba 1993; van der Knaap and Valk 1988).

10.9 Scope for Stem Cell Therapeutics in Neuronal and Neuromuscular Disorders

Neuronal and neuromuscular disorders neither come in one flavor, nor do they manifest at the same time and with the same intensity. For example, diseases or disorders affecting the nervous system can be categorized as either neurodevelopmental or neurodegenerative disorders, caused either due to genetic or non-genetic causes like environmental, nutritional, etc. In neurodevelopmental disorders, the nervous system does not go through normal spatio-temporal differentiation steps which manifest later as behavioral, neurological, or neuromuscular disorders like autism spectrum disorders, Down syndrome, etc. Neurodevelopmental disorders can either be degenerative in nature as in the case of SMA, or non-degenerative as in cerebral palsy, stroke, traumatic brain injury, etc. Neurodegenerative disorders manifest once the nervous system is formed and are characterized by progressive, chronic neurodegeneration with behavioral or motor deficits as exemplified by Alzheimer disease, Huntington chorea, Parkinson disease, etc. (Gupta and Kulhara 2010; Coppede et al. 2006; Daniels 2006; Hardy and Gwinn-Hardy 1998; Price et al. 1998).

While it is almost impossible to correct neurodevelopmental disorders, therapeutic interventions can be envisaged for degenerative disorders, though may not be an outright and actual cure by employing conventional therapeutics. Almost all classical therapeutic approaches, from medication, electrotherapy, radiotherapy, physiotherapy to exercise, excepting curable infections whose effects do not last long, are just palliative in nature. For example, cerebral palsy, a motor function anomaly resulting from abnormality of the immature brain, can only be monitored, but there exists no cure (Garne et al. 2008; Jarvis et al. 2003, 2006; Blair and Watson 2006). Supplementation of a missing factor, for example insulin in diabetes, or relieving pain and suffering, or getting rid of an infectious agent through classical medication or surgical intervention is "relatively" easier, compared to pathological situations wherein the very presence of specific cell or tissue types in structural and functional coordination with the rest of the body is missing. For example, in Hirschsprung disease, which results in pediatric dysmotility due to congenital defect of the ENS wherein a segment of bowel always remains totally aganglionic (Lake and Heuckeroth 2013; Gershon and Ratcliffe 2004; Gariepy 2004; Newgreen and Young 2002a, b; Amiel and Lyonnet 2001; Parisi and Kapur 2000). Neither medication can replace missing ganglia nor ENS is a simple system that can be honed down by classical therapeutics, as it controls and coordinates complex sets of behavior patterns, involving all classes of CNS neurotransmitters in its coordination between bowel and brain.

Many other degenerative diseases, whether congenital or not, also fall into the category of Hirschsprung disease in the sense that progressive deterioration culminates in the loss of neurons and muscles. For example, SMA, ALS, various muscular dystrophies, etc., which either have dying or dead neurons/muscle for which available treatments are at best aimed to limit further deterioration in the health condition. Other disorders which need not necessarily involve loss of neurons may also not respond "effectively" to classical therapeutics. Diseases belonging to the latter category are Alzheimer disease, Huntington chorea, schizophrenia, brain tumor, etc., for which medications are given only to "manage" the disorders, without a cure (Duncan and Valenzuela 2017; Scaglia et al. 2015; Bargiela et al. 2015; Johnson et al. 2014; Fan et al. 2014; Lax et al. 2012; Barker et al. 2012; Laing 2012; Chen et al. 2012; Gerlinger et al. 2012; Ross and Tabrizi 2010; Garne et al. 2008; Javitt et al. 2008; Reif et al. 2006; Delbeuck et al. 2003; Behin et al. 2003; Sullivan et al. 2003; Wong and Van Tol 2003; Holland 2000; van der Knaap and Valk 1988). Given the diversity of diseases and disorders due to wide ranging causes, treatments can be found, not always a cure, only when the causative agent and the underlying physiological mechanisms are known. In the case of known genetic disorders, the CRISPR technology, which is in its infancy, can be used to correct a genetic anomaly in a patient if abnormal cells responsible for a given disease or a disorder still persist without getting irretrievably lost due to degeneration, provided such cells are easily accessible and amenable for corrections (Terns and Terns 2014; Mali et al. 2013). In contrast, one corrective therapy that in most cases overcomes the limitations of conventional therapeutics is regenerative therapeutics based on stem cell biology, which may also employ corrective CRISPR technology if the disorder is due to a genetic error. Stem cell-based regenerative technology has the potential, both to correct and replace lost cell and tissue types, with a few limitations as with any technology (discussed separately). However, it is worth noting that, in contrast to neurodegenerative disorders like Parkinson disease wherein specific populations are lost and specific stem cell-based replacement strategies can be envisaged, cells damaged or lost after stroke, traumatic brain injury, etc, are not limited to any specific neural or non-neural cell type, thus complicating stem cell-directed restructuring of neural circuitry in these cases.

To be an effective treatment technology, it is imperative to understand the biology of stem cells, genome engineering technologies, morphogenesis, physiology, homeostasis/morphostasis, and pathology of the concerned cell or tissue in question. Interestingly, the mechanisms and gene networks involved in morphogenesis, like the Hippo pathway, are also involved in stem cell-mediated morphostasis of any given cell or tissue type (Redondo et al. 2017; Turinetto et al. 2017; Narayanan et al. 2017; She et al. 2017; Yu et al. 2015; Hiemer and Varelas 2013; Ramos and Camargo 2012; Morrison and Kimble 2006). Given this recapitulation of morphogenesis during morphostasis, learning from the basic mechanisms of development and homeostasis will pave way for better engineering in stem cell therapeutics, in general, and neuromuscular and neuronal disorders, in particular (discussed under "Morphostasis recapitulates Morphogenesis").

10.10 Embryonic, Adult and Pluripotent Stem Cells

While contemplating stem cell therapeutics, one encounters three types of stem cells, with differing potencies to give rise to various cell and tissue types. As discussed under "Development and Differentiation of the Nervous and Neuromuscular Systems", only cells that elaborate a fully differentiated body plan are the totipotent embryonic cells. With appropriate cellular and molecular interventions, one can harvest these ES cells and guide them to differentiate into cell types of one's choice for transplantation therapy. As the ES cells are derived from a genetically alien donor embryo, there exists a possibility that differentiated cells/tissues derived from such an embryo might get rejected by the recipient's immune system. Under such a situation, and wherever biology permits, one would employ patient's own adult stem cells that were set aside during development for future use in tissue homeostasis.

Though the use of adult stem cells circumvents the immunological challenges mounted against foreign ES cells, their limited potency to give rise to various cell and tissue types is a clear bottleneck in therapeutics. In addition to this bottleneck, possible depletion of adult stem cells either due to the natural course of aging, infection, or injury also serves as a disadvantage (Mendelsohn and Larrick 2015; Boyette and Tuan 2014; Signer and Morrison 2013; Rosengardten et al. 2011; Piccin and Morshead 2010; Sharpless and DePinho 2007; Reif et al. 2006; Dahlke et al. 2002). However, these limitations that one encounters with embryonic and adult stem cells have been overcome by the introduction of iPSCs (Lloyd-Lewis et al. 2017; Lu and Zhang 2015; De Los Angeles et al. 2015; Morgani and Brickman 2014; Ishiuchi and Torres-Padilla 2013; Maherali and Hochedlinger 2008; Weissman 2004). The ability to induce pluripotency with the guided expression of a specific set of transcription factors paved way for development of personalized iPSCs, and cells and tissues derived from them, thus overcoming the limitations of foreign embryonic and adult stem cells (Lin et al. 2017; Wang et al. 2017a; Hillger et al. 2017;

Genova et al. 2018; Pandya et al. 2017; Li et al. 2014, 2016a, b; Bassuk et al. 2016; Liu and Deng 2016; Jang et al. 2012; Okita et al. 2007; Maherali et al. 2007; Wernig et al. 2007; Takahashi and Yamanaka 2006; Tokuzawa et al. 2003). Various attempts have been made to treat neuromuscular disorders by employing these iPSCs in multiple animal models (Orbay and Mizuno 2013).

10.11 Stem Cell Transplantation, Delivery, Homing, and Engraftment

One of the main driving forces for successful stem cell therapeutics upon delivery is the ability of stem cells to colonize or home the damaged tissue areas which is often referred to as stem cell homing (Jackson et al. 2010; Lamfers et al. 2009; Blocklet et al. 2006; Torrente et al. 2003, 2006; Moore et al. 2004; Ji et al. 2004; Imitola et al. 2004; Lee et al. 2003). Subsequent to homing the damaged tissue, stem cells get engrafted to the tissue for repairing the damage (Hall et al. 2017; Lala-Tabbert et al. 2016; Zhang et al. 2015; Anderson et al. 2011; Tamaki et al. 2002; Lacorazza et al. 1996). Though stem cell transplantation/delivery may not be a critical factor while regenerating a tissue or an organ on biological scaffolds or by 3D printing in vitro for subsequent transplantation, it is a matter of importance if one were to correct a cell or a tissue type in vivo.

In principle, there are two major methods of delivering stem cells to the body, local delivery to the tissue and delivery through systemic circulation. Local delivery can be achieved by either employing biological scaffolds or by local injection to the tissue in question. Though the delivery of stem cells to peripheral cells and tissues by local injection is relatively simple and easy, deeper and complex tissues and organs require specialized treatments. While targeted injection of stem cells deep into the tissue in question could be one way to address this issue, it becomes a tedious task to embark when more than one tissue or muscle is affected. This drawback can be circumvented by delivering stem cells via systemic circulation. Systemic delivery is achieved by the vascular route, either by IV or IA modes (Leibacher and Henschler 2016; Arthur et al. 2009). While systemic circulation can be employed for delivering stem cells to the tissues like musculature, the same cannot be the means for their delivery into the CNS due to the BBB. Though, targeted delivery of stem cells to brain by invasive methods can circumvent BBB, novel non-invasive methods in model organisms and humans like intranasal delivery can effectively bypass the BBB mainly via olfactory pathways (Balyasnikova et al. 2014; van Velthoven et al. 2013; Wu et al. 2013; Bossolasco et al. 2012; Danielyan et al. 2009, 2011). Upon delivery to brain and given their ability to home in damaged sites, stem cells help heal affected areas of the brain.

Whenever stem cells are transplanted, irrespective of the delivery methods employed, they are often exposed to a hostile tissue environment that would potentially lead to their apoptosis with low level of cell survival (Parr et al. 2007; Kallur et al. 2006; Bakshi et al. 2005). In order to employ stem cell transplantation successfully as an effective therapeutic option, it is imperative that stem cell survival is ensured and maintained. The stem cell niche in vivo acts both as a buffering system and a microenvironment to receive and interpret a host of cellular and biological signals to regulate tissue regeneration according to biological demand or against tissue injury (Discher et al. 2009; Lane et al. 2009). In a therapeutic setting and in the absence of such a niche, stem cell survival gets compromised, thus negatively influencing recovery. Hence, a successful stem cell therapeutic intervention necessitates creating microenvironments/niches conducive for stem cell survival that helps integrate transplanted stem cells with the target tissue. This type of niche for stem cells is created by using biomaterial scaffolds. These scaffolds are designed to have biocompatible and mechanical properties relevant to stem cells in question along with providing cell adhesion and survival factors. Biomaterials have the propensity to form hydrogels like collagen and agarose, and have been used as scaffolds to create artificial niche for stem cells; biomaterial scaffolds fine-tuned for stem cells targeted to the muscle and the nervous system have been developed for better success in therapeutics (Lane et al. 2014; Kim et al. 2012; Wagers 2012; Peerani and Zandstra 2010; Adams et al. 2007).

10.12 Major Stem Cell Therapeutic Approaches

Based on the type of disorder and the techniques used, one could classify stem cell therapeutics for disorders of nerve and muscle into following three categories:

1. Treatment with stem cell-derived extracellular vesicles and mitochondria
2. Stem cell-based cell and tissue replacement
3. Modeling and drug screening

Screen for drugs against diseases of muscles and nerves is done by using stem cell models carrying the disease-causing mutations such that the drug so selected will work best in the patient. On the contrary, other two treatment methods directly involving stem cells may need to go through editing the underlying patient-specific genetic anomaly using technologies like CRISPR, for an effective cure.

10.13 Stem Cell-Derived Extracellular Vesicles and Mitochondria

A new approach to stem cell therapeutics involves, rather than, or in addition to the conventional stem cell replacement, the EVs released from stem cells. Intercellular transfer of EVs is an established mechanism of cell communication and represents a means to transfer or exchange macromolecules between cells (Lai et al. 2015;

Maumus et al. 2013). Upon secretion, vesicles interact with target cells to deliver their cargo of biological relevance (Lai et al. 2015). This property of stem cells to transfer their cargo to target cells can be employed for therapeutic purposes. Disorders most appropriate for treatments with EVs are LSDs and mitochondrial neuropathies. LSDs are due to the defective activity of lysosomal hydrolases caused by pathogenic mutations in their encoding genes or in other components critical for lysosomal function. Though the functions of lysosomes are fundamental in every cell type and the consequences of lysosomal impairment are also widespread, majority of LSD patients suffer from CNS anomalies with varied severity. Since EVs like exosomes and microvesicles are known to transfer long range biological information and as microvesicles can cross the BBB due to their small size, they have potential therapeutic applications in LSDs, ALS, Alzheimer and Parkinson diseases (Lee et al. 2016; Yuyama et al. 2015; Jarmalaviciute et al. 2015; Farinazzo et al. 2015; Katsuda et al. 2013).

In addition to the classical delivery of cargo by EVs from stem cells to the defective cells, the transfer of mitochondria between cells is important, especially with regard to mitochondrial neuropathies and myopathies. Mitochondrial disorders can be of two types: those with mutations in the nuclear genes required for mitochondrial biology and those that are inherent to the mitochondrial genome. Though the former can be corrected by employing recent genome editing technologies like CRISPR, the same cannot be said to be of much value for inherent mitochondrial pathologies. Since transfer of mitochondria from stem cells to epithelial cells of the airway and cornea is known to protect against acute injuries to lung and cornea (Dan Jiang et al. 2016; Islam et al. 2012), this method is of potential value for treating mitochondrial neuropathies and myopathies (Pfeffer and Chinnery 2013; Pareyson et al. 2013; Finsterer 2005, 2011; Hassani et al. 2010). In addition, stem cell technologies have been evolved to generate mutation-free lines of stem cells for mitochondrial therapeutics. Nuclei of skin cells isolated from patients diagnosed with severe mitochondrial disorders like mitochondrial encephalomyopathy and Leigh Syndrome, which affect brain and muscles, were injected into a donor egg cell harboring healthy mitochondria (Ma et al. 2015). Non-mutant mitochondrial stem cell lines so established could be made to differentiate into cell type of relevance for therapeutic purpose to generate disease-free heart, brain, or the muscle. The ability to transform patient-derived diseased cells into pluripotent stem cells having healthy mitochondrial DNA opens up a possibility to envisage stem cell-mediated delivery of mitochondria to affected tissues or even replacing the whole tissue.

As specific control of gene expression via RNAi and miRNAs is known to mediate migration and differentiation of stem cells, stem cells are also considered as delivery vectors for RNAi in CNS diseases in the same way as they are used for delivering EVs and mitochondria (Hou et al. 2017).

10.14 Choice of Stem Cells, Stem Cell Plasticity and Trans-differentiation

NPSCs and satellite cells are the stem cells that contribute towards neural and muscle regeneration, respectively and have immense therapeutic potential. However, given the diversity of neural subtypes and the limited ability to harvest and engineer them, additional sources of stem cells for treating the disorders of nerve and muscle have been explored. These efforts have exploited the developmental phenomenon of trans-differentiation. Trans-differentiation is the differentiation of one type of cell in a tissue or an organ into a different cell type, which is characteristic of another tissue or organ, without evolving through a pluripotent cell state. A best example for trans-differentiation is the one seen in the liver and the pancreas. These two organs comprising multiple cell types develop from the adjoining regions of the endodermal epithelium and trans-differentiation of pancreas to liver and vice versa have been recorded (Deutsch et al. 2001; Wells and Melton 1999).

Similar trans-differentiation events are seen in non-neural and non-muscle tissues. For example, MEFs have been induced to trans-differentiate into NSCs and these differentiated cells could significantly improve the motor and sensory functions of rats with MCAO (Yao et al. 2015). Further, efforts have also been made to generate cells resembling neural stem cells from stromal cells of the adult human bone marrow (Hermann et al. 2004). Similarly, trans-differentiation property has been explored in stem cell therapy as an option to treat neuronal disorders (Alexiades and Mckhann II 2017; Salehi et al. 2016; Hermann and Storch 2013; Kmet et al. 2013; Thier et al. 2012; Shi et al. 2012a, b; Hermann et al. 2006a, b; Krabbe et al. 2005; Li et al. 2005).

Non-muscle cells like pericytes on small blood vessels can transform to make both muscle fibers and muscle stem cells during normal development (Dellavalle et al. 2011). Further, as it has been shown that pericytes can contribute to muscle development and repair both during normal muscle growth and in extreme situations, an understanding of normal neural and neuromuscular development and their underlying mechanisms greatly contributes to fine-tuning stem cell therapeutics. In similar lines, it has also been shown that bone marrow transplantation attenuated the myopathic phenotype in a mouse model of SMA, though more likely through trophic factors rather than their capacity to replace or fuse to mutant myofibers (Salah-Mohellibi et al. 2006). Similar to pericytes, SP cells, CD133+ cells, MABs, PICs, and β4 integrin+ cells are also known to acquire myogenic properties (Pannerec et al. 2012). Interestingly, even adult neural stem cells are shown to trans-differentiate into myogenic cells upon transplantation into regenerating adult skeletal muscle (Galli et al. 2000). Further, adult mesenchymal stem cells, which are found in almost all tissue types and play essential roles in tissue homeostasis, renewal, and repair, seem to be highly plastic with great trans-differentiation potential (Sousa et al. 2014; Tuan et al. 2003). Given their ability to repair tissue damage, mesenchymal stem cells from bone marrow and adipose tissue are two of the widely used mesenchymal stem cell sources for musculo-skeletal therapeutics (de Bakker et al. 2013; Mafi et al. 2011; Gutierrez-Nibeyro 2011; Fortier and Travis 2011; Frisbie and Smith 2010; Neupane et al. 2008).

10.15 Stem Cells in Neurological and Muscle Disorders

As discussed earlier, neurological disorders can be looked in various ways. For example, one could distinguish neurological disorders affecting sensory and motor neurons from those that affect neurons with differing neurotransmitters. According to the type of neurons affected, stem cells have to be guided to differentiate into specific neurons of choice. However, in cases wherein sensory or motor neurons are irreversibly lost, it has not yet been possible to connect a stem cell-derived motor or sensory neuron from its origin to the target tissue and neither do we have a technology to lure a stem cell induced neuron to extend its axon all the way to the specific target tissue. On the contrary, two types of regenerative therapeutics could be envisaged for the replacement of cells lost in the CNS. While pharmacological activation of endogenous stem cells is one such method (Cooke et al. 2011; Obermair et al. 2008), exogenous stem cell transplantation is the other one. The latter is used to replace lost cells at the site of injury or repair the damaged or defective cells as in lysosomal storage disorders, mitochondrial myopathies and neuropathies or for remyelinating neurons. Alternatively, exogenously transplanted stem cells are employed to provide local milieu to induce existing adult stem cells to regenerate by providing trophic signals (Ruff and Fehlings 2010; Tetzlaff et al. 2011).

NPC-based therapies have been attempted for neuroinflammatory disorders in model systems. Chronic degeneration of the CNS due to inflammation was shown to be efficiently protected by transplanted NPCs both in rodents and non-human primates (Pluchino et al. 2003, 2009; Bottai et al. 2008; Ben-Hur et al. 2003; Chu et al. 2003). The other disorder with neuromuscular involvement in the context of stem cell therapeutics is SMA, the second most frequent genetic disorder. Spinal cord derived neural stem cells, neural stem cells derived from pluripotent stem cells, murine ES cells, and embryonic stem cell lines have all ameliorated the disease phenotype in the mouse model of SMA (Corti et al. 2008, 2010). Spinal cord-derived neural stem cells have not only contributed to therapeutic improvements seen in SMA, but also prevented skeletal muscle atrophy following peripheral nerve injury, thus extending their possible use in treating similar muscular atrophies (Ruven et al. 2017).

Unlike other neurological disorders, clinical trials using stem cells have not shown considerable improvements in Alzheimer disease patients (Kang et al. 2016b). Though various possibilities can be envisaged, stem cell therapeutics for diseases like autism spectrum disorders, schizophrenia, neuropathic pain, polyglutamine disorders, Alzheimer and Parkinson diseases need further innovations in technology and a better understanding of developmental biology (Shetty and Bates 2016; Ichim et al. 2007). Equally promising is the ability of MSCs to seek out damaged cells and their tumor tropism. The MSCs from adipose tissue could be used for delivery of adjuvant therapies against gliomas (Pendleton et al. 2013). Following this pioneering work, it has recently been shown to work with reprogrammed stem cells in significantly reducing mouse glioma (Bago et al. 2017).

Muscle disorders are of varying types from post-polio syndrome to DMD. On the contrary, skeletal muscle injuries are common in traumatic scenarios like war, road- or job-related events. Though skeletal muscles have good regenerative abilities, the extent to which they can repair themselves, both structurally and functionally, may be limited under conditions of severe tissue loss and would finally result in the build-up of non-contractile fibrous tissue scar. Further, RICE based methods and surgical procedures currently in vogue are known to be not effective against severe loss of muscle tissue, thus necessitating novel therapeutic avenues (Jarvinen et al. 2005, 2007, 2014). Whether due to infection, genetic loss of function or due to injury, one of the therapeutic avenues being employed to regenerate a healthy muscle involves the application of stem cells residing in muscle and non-muscle stem cells (Maclean et al. 2012; Tedesco et al. 2010).

Satellite cells are activated upon muscle damage leading to their proliferation and formation of MPCs, which help in rebuilding the damaged muscle, though some of the stem cells self-renew and replenish the quiescent satellite cells (Rossi et al. 2011; Collins et al. 2005). However, due to the decreased cell availability of the postnatal satellite cells and given the propensity of native tissue skeletal MDSCs to dedifferentiate (Page et al. 2011; Mizuno et al. 2002; Yaffe 1968), alternative sources of stem cells either by generating iPSCs or by exploting the trans-differentiation property, new modes of stem cell therapeutics for the skeletal muscle need to be explored.

MSCs from different sources are shown to be capable of differentiation into myogenic lineages expressing markers unique to skeletal muscle cells (Hodgson et al. 2018; Lorant et al. 2018). One of the major muscle disorders wherein MSCs are successfully tested is DMD, the most studied and a widespread genetic disorder carrying mutations in the *dystrophin (DYS)* gene, thus projecting possible therapeutic options for other muscle disorders (Pisciotta et al. 2015; Rajput et al. 2015; Fujita et al. 2015; Sohn et al. 2015; Gonçalves et al. 2006; De Bari et al. 2003; Corti et al. 2002). Attempts were also made to rescue a mouse model lacking the *dystrophin* gene (referred to as *mdx* mouse) by introducing the entire human *dystrophin* gene through the human artificial chromosomes (DYS-HAC) into blood vessel-associated stem cells (originally named MABs). Upon transplanting DYS-HAC carrying MABs back into the immunosuppressed *mdx* animals, it was found that the transplanted cells not only expressed the human *dystrophin* gene but also proliferated and differentiated into muscle cells. Consequently, the mutant animals showed considerable improvements in their overall behavioral performance and morphology (Tedesco et al. 2011). However, with the advancements in genome editing technologies like CRISPR, correcting mutant stem cells and reintroducing them back into the animal as a corrective technology for muscular dystrophies has not only become more precise, but also made simple to achieve success (Pini et al. 2017). MSCs based therapies have also been tried for SMA as these patients have disorders of both the nerve and the muscle. Three children with SMA type 1 had been subjected to multiple intrathecal and intravenous infusions of MSCs have met with some degree of success in alleviating the symptoms associated with the disorder (Villanova and Bach 2015).

Gut neuromuscular diseases fall into a class of their own. As detailed earlier, Hirschsprung disease results in pediatric dysmotility due to congenital defect of the ENS with a segment of bowel always remaining totally aganglionic. As neither medication can replace the missing ganglia nor ENS is a simple system that can be honed down by classical therapeutics, identification of ENS stem cells in fetal and adult mammals throws open the possibility of autologous stem cell therapy for intestinal motility disorders, including Hirschsprung disease (Metzger et al. 2009a, b; Bixby et al. 2002).

10.16 Stem Cells for Modeling and Drug Screening

High throughput drug screening technologies often involve immortalized cell lines that do not reflect the normal cells of the body, let alone that of the patients. On the contrary, availability of stem cells, which can even be made disease or patient specific using iPSC technology, opened up opportunities to develop more specific, personalized and effective drug screening platforms (Ebert and Svendsen 2010). Numerous patient derived iPSC disease models have been generated that range from Duchenne and Becker muscular dystrophy, Huntington chorea, ALS, SMA, etc. (Kang et al. 2016a; Schwartz et al. 2015; Chal et al. 2015; Heilker et al. 2014; Ko and Gelb 2014; Alcendor et al. 2013; Mackay-Sim 2013; Kim and Lee 2013; Robinton and Daley 2012; Egawa et al. 2012; Liu and Zhang 2010; Park et al. 2008). These drug screening methods open up both general and personalized therapeutic options for diseases of the muscle and nerve that are more precise than the conventional medications.

10.17 Limitations and Future of Stem Cell Therapeutics

If not now, but certainly in the distant future, one may be able to correct or replace almost every neuron of the nervous system by using stem cell therapeutics. Correcting any biochemically or hormonally relevant properties of brain control mechanism would not be too difficult. However, one of the major hurdles in the way of neuronal stem cell therapeutics are treating diseases and disorders linked to traits in human beings, such as knowledge, language, intelligence, experience, and memory of the past. Our subjective experiences of the past, involving feelings and emotions, cannot be brought into newly replaced stem cell-derived neurons. Such stem cell-derived neurons can certainly account for making new neural networks and experiences, but all the previous experiences and knowledge of an individual are lost forever. Hence, lost psychological memory involving experiences, feelings, personal identity, etc. cannot be fixed just by making new memory forming cells as they are irretrievably lost forever due to specific neurological/neurodegenerative disorders. Though the fundamental aspects of manifest memory are compromised, the underlying causative principle that is compromised in such a scenario is worth a mention. For example, if the visual, linguistic, and taste attributes of an apple are stored in different parts of the brain, then repairing and replacing neurons in any one of the affected areas by employing stem cell therapeutics could potentially "delete" all the previous information required to recognize any specific attribute of an apple, or any other object for that matter. As we do not have a clear molecular understanding of how exactly memory and experiences of the past are stored, integrated, and retrieved in a meaningful way, there is no way to create a backup of our experiences of the past in order to be copied into new stem cell-derived neurons in the brain. In light of these issues, stem cell therapeutics can potentially deprive a person of his identity, which the patient had lost due to dementia or any other disorder affecting individual identity to begin with, thus creating fresh bioethical concerns. For example, though the experience of the past cannot be retrieved if the neurons responsible for memory storage are lost forever, impaired acquisition of new memory due to loss of basal forebrain cholinergic neurons that project to the hippocampus and after intrinsic hippocampal disease can be corrected in experimental models using iPSC-derived cholinergic neuronal replacement (Goldman 2016; Liu et al. 2013). Hence, it is a huge challenge for any of the therapeutic options currently in vogue to restore the intellectual, behavioral, and social experiences of the past and remains as a big challenge for the future.

Given the genetic variations, ES cells from a donor embryo could be treated as antigenic/foreign by the patient immune system posing immunological problems and are likely to be rejected subsequent to transplantation. This drawback can be circumvented by using patient derived iPSCs. However, unlike embryonic or adult stem cells, iPSCs are derived from the differentiated cells of an adult, which through their life in a patient could have accumulated mutations. Tissues and organs derived from such pluripotent cells would suffer from mutation load and would not be appropriate, both clinically and ethically to be used in therapeutics as doing so would be counterproductive (Vitale et al. 2017). Noteworthy, sourcing of embryonic or adult stem cells should not pose a serious challenge in therapeutics due to their robust ability to repair DNA damage, which prevents accumulation of mutations in their differentiated adult structures. However, the same cannot be said of cells differentiated from the patient-derived pluripotent stem cells. How to blend the positive properties of ES cells and iPSCs for a successful therapy is a challenge that is yet to be addressed.

Timing of therapeutic intervention using stem cell technologies is a very important criterion for neuromuscular disorders. For example, given the interdependence of the two synaptic partners, nerve and the muscle, loss of one would lead to gradual loss of the other. In such a situation, correcting one of the synaptic partners will not necessarily improve the pathological and behavioral outcomes and, in fact, further complicate the available therapeutic options. With a better understanding of the biology of development and differentiation, advent of new stem cell therapeutic avenues and drugs would pave way for better healthcare and treatment. Further, directing a stem cell to develop and differentiate into a specific tissue or organ of choice is critically dependent on clues obtained by studying the genetic and molecular basis of development and differentiation. More importantly, our limited understanding of developmental biology has not served us better in dealing with severe motoneuron disorders or disorders affecting memory and intelligence. For example, if and when a motoneuron is lost due to a disease or a disorder, connecting a stem cell-derived motoneuron from the spinal cord to the target skeletal muscle has not yet been possible. Neither do we have a technology to lure a stem cell-induced motor neuron in the spinal cord to extend its axons all the way to specific target skeletal muscles. Hence, a better understanding of developmental biology greatly facilitates harnessing its power in effective stem cell-based therapeutics. A lot is achieved and yet a lot more is desired.

Acknowledgements: Authors duly acknowledge the help and encouragement from Dr. D. Y. Patil Vidyapeeth, Pimpri, Pune, Maharashtra, India.

References

Adams GB, Martin RP, Alley IR et al (2007) Therapeutic targeting of a stem cell niche. Nat Biotechnol 25:238–243
Al Bekairy AM, Al Harbi S, Alkatheri AM et al (2014) Bacterial Meningitis: an update review. Afr J Pharm Pharmacol 8:469–478
Alcendor DJ, Block FE III, Cliffel DE et al (2013) Neurovascular unit on a chip: implications for translational applications. Stem Cell Res Ther 4(Suppl 1):S18. https://doi.org/10.1186/scrt379
Al-Chalabi A, Hardiman O (2013) The epidemiology of ALS: a conspiracy of genes, environment and time. Nat Rev Neurol 9:617–628
Alexiades NG, Mckhann GM II (2017) Getting more skin in the game: using human neural stem cells transdifferentiated from fibroblasts to target malignant gliomas. Neurosurgery 81:N36–N37
Amiel J, Lyonnet S (2001) Hirschsprung disease, associated syndromes, and genetics: a review. J Med Genet 38:729–739
Andersen PM, Al-Chalabi A (2011) Clinical genetics of amyotrophic lateral sclerosis: what do we really know? Nat Rev Neurol 7:603–615
Anderson DJ (1993) Molecular control of cell fate in the neural crest: the sympathoadrenal lineage. Annu Rev Neurosci 16:129–158
Anderson DJ (1997) Cellular and molecular biology of neural crest cell lineage determination. Trends Genet 13:276–280
Anderson AJ, Haus DL, Hooshmand MJ et al (2011) Achieving stable human stem cell engraftment and survival in the CNS: is the future of regenerative medicine immunodeficient? Regen Med 6:367–406
Appel B, Korzh V, Glasgow E et al (1995) Motoneuron fate specification revealed by patterned LIM homeobox gene expression in embryonic zebrafish. Development 121:4117–4125
Arnold HH, Braun T (2000) Genetics of muscle determination and development. Curr Top Dev Biol 48:129–164
Arthur A, Zannettino A, Gronthos S (2009) The therapeutic applications of multipotential mesenchymal/stromal stem cells in skeletal tissue repair. J Cell Physiol 218:237–245
Asa SL, Ezzat S (2002) The pathogenesis of pituitary tumours. Nat Rev Cancer 2:836–849

Babin PJ, Goizet C, Raldua D (2014) Zebrafish models of human motor neuron diseases: advantages and limitations. Prog Neurobiol 118:36–58
Bago JR, Okolie O, Dumitru R et al (2017) Tumor-homing cytotoxic human induced neural stem cells for cancer therapy. Sci Transl Med 9(375):pii: eaah6510. https://doi.org/10.1126/scitranslmed.aah6510
Baj A, Colombo M, Headley JL et al (2015) Post-poliomyelitis syndrome as a possible viral disease. Int J Infect Dis 35:107–116
Bakshi A, Keck CA, Koshkin VS et al (2005) Caspase-mediated cell death predominates following engraftment of neural progenitor cells into traumatically injured rat brain. Brain Res 1065:8–19
Balyasnikova IV, Prasol MS, Ferguson SD et al (2014) Intranasal delivery of mesenchymal stem cells significantly extends survival of irradiated mice with experimental brain tumors. Mol Ther 22:140–148
Bargiela D, Shanmugarajah P, Lo C et al (2015) Mitochondrial pathology in progressive cerebellar ataxia. Cerebellum Ataxias 2:16
Barker RA, Cicchetti F, Neal MJ (2012) Neuroanatomy and neuroscience at a glance. Wiley-Blackwell, New Jersey
Bartholow R (1982) Medical electricity: a practical treatise on the applications of electricity to medicine and surgery. Henry C. Lea's Son & Co., Philadelphia
Bartt R (2012) Acute bacterial and viral meningitis. Continuum Lifelong Learning Neurol 18:1255–1270
Baser A, Skabkin M, Martin-Villalba A (2017) Neural stem cell activation and the role of protein synthesis. Brain Plast 3:27–41
Bassuk AG, Zheng A, Li Y et al (2016) Precision medicine: genetic repair of retinitis pigmentosa in patient-derived stem cells. Sci Rep 6:19969. https://doi.org/10.1038/srep19969
Behin A, Hoang-Xuan K, Carpentier AF et al (2003) Primary brain tumours in adults. Lancet 361:323–331
Belgacem YH, Hamilton AM, Shim S et al (2016) The many hats of Sonic Hedgehog signaling in nervous system development and disease. J Dev Biol 4:35. https://doi.org/10.3390/jdb4040035
Ben-Hur T, Einstein O, Mizrachi-Kol R et al (2003) Transplanted multipotential neural precursor cells migrate into the inflamed white matter in response to experimental autoimmune encephalomyelitis. Glia 41:73–80
Bentzinger CF, Wang YX, Rudnicki MA (2012) Building muscle: molecular regulation of myogenesis. Cold Spring Harb Perspect Biol 4. https://doi.org/10.1101/cshperspect.a008342
Bergsland M, Ramskold D, Zaouter C et al (2011) Sequentially acting Sox transcription factors in neural lineage development. Genes Dev 25:2453–2464
Bernstein JJ, Mary EB (1971) Axonal regeneration and formation of synapses proximal to the site of lesion following hemisection of the rat spinal cord. Exp Neurol 30:336–351
Bertini E, D'Amico A, Gualandi F et al (2011) Congenital muscular dystrophies: a brief review. Semin Pediatr Neurol 18:277–288
Bertrand N, Castro DS, Guillemot F (2002) Proneural genes and the specification of neural cell types. Nat Rev Neurosci 3:517–530
Birchmeier C, Brohman H (2000) Genes that control the development of migrating muscle precursor cells. Curr Opin Cell Biol 12:725–730
Bixby S, Kruger GM, Mosher JT et al (2002) Cellintrinsic differences between stem cells from different regions of the peripheral nervous system regulate the generation of neural diversity. Neuron 35:643–656
Blair E, Watson L (2006) Epidemiology of cerebral palsy. Semin Fetal Neonatal Med 11:117–125
Blau HM, Brazelton TR, Weimann JM (2001) The evolving concept of a stem cell: entity or function. Cell 105:829–841
Blau HM, Cosgrove BD, Ho AT (2015) The central role of muscle stem cells in regenerative failure with aging. Nat Med 21:854–862
Blocklet D, Toungouz M, Berkenboom G et al (2006) Myocardial homing of nonmobilized peripheral-blood CD34+ cells after intracoronary injection. Stem Cells 24:333–336

Bossolasco P, Cova L, Levandis G et al (2012) Noninvasive near-infrared live imaging of human adult mesenchymal stem cells transplanted in a rodent model of Parkinson's disease. Int J Nanomed 7:435–447
Bottai D, Madaschi L, Di Giulio AM et al (2008) Viability-dependent promoting action of adult neural precursors in spinal cord injury. Mol Med 14:634–644
Boyette LB, Tuan RS (2014) Adult stem cells and diseases of aging. J Clin Med 3:88–134
Brack AS, Munoz-Canoves P (2016) The ins and outs of muscle stem cell aging. Skelet Muscle 6:1
Breunig JJ, Levy R, Antonuk CD et al (2015) Ets factors regulate neural stem cell depletion and gliogenesis in Ras pathway glioma. Cell Rep 12:258–271
Brzustowicz LM, Lehner T, Castilla LH et al (1990) Genetic mapping of chronic childhood-onset spinal muscular atrophy to chromosome 5q11.2-13.3. Nature 344:540–541
Buckingham M (2001) Skeletal muscle formation in vertebrates. Curr Opin Genet Dev 11:440–448
Budnik V (1996) Synapse maturation and structural plasticity at *Drosophila* neuromuscular junctions. Curr Opin Neurobiol 6:858–867
Burkin DJ, Kaufman SJ (1999) The alpha7beta1 integrin in muscle development and disease. Cell Tissue Res 296:183–190
Byrne GJ (2005) Pharmacological treatment of behavioural problems in dementia. Aust Prescr 28:67–70
Carpio A, Romo ML, Parkhouse RM et al (2016) Parasitic diseases of the central nervous system: lessons for clinicians and policy makers. Expert Rev Neurother 16:401–414
Chakkalakal JV, Jones KM, Basson MA et al (2012) The aged niche disrupts muscle stem cell quiescence. Nature 490:355–360
Chal J, Oginuma M, Al Tanoury Z et al (2015) Differentiation of pluripotent stem cells to muscle fiber to model Duchenne muscular dystrophy. Nat Biotechnol 33:962–969
Chan AW, Dominko T, Luetjens CM et al (2000) Clonal propagation of primate offspring by embryo splitting. Science 287:317–319
Chang Q, Balice-Gordon RJ (2000) Highwire, rpm-1, and futsch: balancing synaptic growth and stability. Neuron 26:287–290
Chau CL, Griffith JF (2005) Musculoskeletal infections: ultrasound appearances. Clin Radiol 60:149–159
Chen JN, Fishman MC (2000) Genetics of heart development. Trends Genet 16:383–388
Chen J, Li Y, Yu TS et al (2012) A restricted cell population propagates glioblastoma growth after chemotherapy. Nature 488:522–526
Chien KR (2004) Stem cells: lost in translation. Nature 428:607–608
Chu K, Kim M, Jeong SW et al (2003) Human neural stem cells can migrate, differentiate, and integrate after intravenous transplantation in adult rats with transient forebrain ischemia. Neurosci Lett 343:129–133
Chu-LaGraff Q, Schmid A, Leidel J et al (1995) huckebein specifies aspects of CNS precursor identity required for motorneuron axon pathfinding. Neuron 15:1041–1051
Cockburn N, Pradhan A, Taing MW et al (2017) Oral health impacts of medications used to treat mental illness. J Affect Disord 223:184–193
Collins CA, Olsen I, Zammit PS et al (2005) Stem cell function, self-renewal, and behavioral heterogeneity of cells from the adult muscle satellite cell niche. Cell 122:289–301
Cooke MJ, Wang Y, Morshead CM et al (2011) Controlled epi-cortical delivery of epidermal growth factor for the stimulation of endogenous neural stem cell proliferation in stroke-injured brain. Biomaterials 32:5688–5697
Coppede F, Mancuso M, Siciliano G et al (2006) Genes and the environment in neurodegeneration. Biosci Rep 26:341–367
Cornelison DD, Wold BJ (1997) Single-cell analysis of regulatory gene expression in quiescent and activated mouse skeletal muscle satellite cells. Dev Biol 191:270–283
Corti S, Strazzer S, Del Bo R et al (2002) A subpopulation of murine bone marrow cells fully differentiates along the myogenic pathway and participates in muscle repair in the mdx dystrophic mouse. Exp Cell Res 277:74–85

Corti S, Nizzardo M, Nardini M et al (2008) Neural stem cell transplantation can ameliorate the phenotype of a mouse model of spinal muscular atrophy. J Clin Invest 118:3316–3330
Corti S, Nizzardo M, Nardini M et al (2010) Embryonic stem cell-derived neural stem cells improve spinal muscular atrophy phenotype in mice. Brain 133:465–481
Crum-Cianflone NF (2008) Bacterial, fungal, parasitic, and viral myositis. Clin Microbiol Rev 21:473–494
Cruza PMR, Palacea J, Beeson D (2014) Congenital myasthenic syndromes and the neuromuscular junction. Curr Opin Neurol 27:566–575
Dahlke MH, Lauth OS, Jager MD et al (2002) In vivo depletion of hematopoietic stem cells in the rat by an anti-CD45 (RT7) antibody. Blood 99:3566–3572
Dahmann C, Oates AC, Brand M (2011) Boundary formation and maintenance in tissue development. Nat Rev Genet 12:43–55
Daniels SK (2006) Neurological disorders affecting oral, pharyngeal swallowing. GI Motility Online. https://doi.org/10.1038/gimo34
Danielyan L, Schäfer R, von Ameln-Mayerhofer A et al (2009) Intranasal delivery of cells to the brain. Eur J Cell Biol 88:315–324
Danielyan L, Schäfer R, von Ameln-Mayerhofer A et al (2011) Therapeutic efficacy of intranasally delivered mesenchymal stem cells in a rat model of Parkinson disease. Rejuvenation Res 14:3–16
Dasen JS, Jessell TM (2009) Hox networks and the origins of motor neuron diversity. Curr Top Dev Biol 88:169–200
Day SJ, Lawrence PA (2000) Measuring dimensions: the regulation of size and shape. Development 127:2977–2987
de Bakker E, Van Ryssen B, De Schauwer C et al (2013) Canine mesenchymal stem cells: state of the art, perspectives as therapy for dogs and as a model for man. Vet Q 33:225–233
De Bari C, Dell'Accio F, Vandenabeele F et al (2003) Skeletal muscle repair by adult human mesenchymal stem cells from synovial membrane. J Cell Biol 160:909–918
De Los Angeles A, Ferrari F, Xi R et al (2015) Hallmarks of pluripotency. Nature 525:469–478
Delbeuck X, Van der Linden M, Collette F (2003) Alzheimer's disease as a disconnection syndrome? Neuropsychol Rev 13:79–92
Dellavalle A, Maroli G, Covarello D et al (2011) Pericytes resident in postnatal skeletal muscle differentiate into muscle fibres and generate satellite cells. Nat Commun 2:499. https://doi.org/10.1038/ncomms1508
Deutsch G, Jung J, Zheng M et al (2001) A bipotential precursor population for pancreas and liver within the embryonic endoderm. Development 128:871–881
Di Bonito M, Glover JC, Studer M (2013) *Hox* genes and region-specific sensorimotor circuit formation in the hindbrain and spinal cord. Dev Dyn 242:1348–1368
Discher DE, Mooney DJ, Zandstra PW (2009) Growth factors, matrices, and forces combine and control stem cells. Science 324:1673–1677
Doherty JT, Lenhart KC, Cameron MV et al (2011) Skeletal muscle differentiation and fusion are regulated by the BAR-containing Rho-GTPase-activating protein (Rho-GAP), GRAF1. J Biol Chem 286:25903–25921
Dun XP, Parkinson DB (2017) Role of Netrin-1 signaling in nerve regeneration. Int J Mol Sci 18:491. https://doi.org/10.3390/ijms18030491
Duncan T, Valenzuela M (2017) Alzheimer's disease, dementia, and stem cell therapy. Stem Cell Res Ther 8:111
Dyck PJ, Kratz KM, Karnes JL et al (1993) The prevalence by staged severity of various types of diabetic neuropathy, retinopathy, and nephropathy in a population-based cohort: the Rochester Diabetic Neuropathy Study. Neurology 43:817–824
Eade KT, Fancher HA, Ridyard MS (2012) Developmental transcriptional networks are required to maintain neuronal subtype identity in the mature nervous system. PLoS Genet 8:e1002501
Ebert AD, Svendsen CN (2010) Human stem cells and drug screening: opportunities and challenges. Nat Rev Drug Discov 9:367–372

Edens BM, Ajroud-Driss S, Ma L et al (2015) Molecular mechanisms and animal models of spinal muscular atrophy. Biochim Biophys Acta 1852:685–692

Egawa N, Kitaoka S, Tsukita K et al (2012) Drug screening for ALS using patient-specific induced pluripotent stem cells. Sci Transl Med 4:145ra104. https://doi.org/10.1126/scitranslmed.3004052

Engel AG, Shen XM, Selcen D et al (2015) Congenital myasthenic syndromes: pathogenesis, diagnosis, and treatment. Lancet Neurol 14:420–434

Engesser-Cesar C, Anderson AJ, Basso DM et al (2005) Voluntary wheel running improves recovery from a moderate spinal cord injury. J Neurotrauma 22:157–171

Enns GM, Huhn SL (2008) Central nervous system therapy for lysosomal storage disorders. Neurosurg Focus 24:E12

Erichsen AK, Koht J, Stray-Pedersen A et al (2009) Prevalence of hereditary ataxia and spastic paraplegia in southeast Norway: a population-based study. Brain 132:1577–1588

Eriksson PS, Perfilieva E, Bjork-Eriksson T et al (1998) Neurogenesis in the adult human hippocampus. Nat Med 4:1313–1317

Fan HC, Ho LI, Chi CS et al (2014) Polyglutamine (PolyQ) diseases: genetics to treatments. Cell Transplant 23:441–458

Farinazzo A, Turano E, Marconi S et al (2015) Murine adipose-derived mesenchymal stromal cell vesicles: in vitro clues for neuroprotective and neuroregenerative approaches. Cytotherapy 17:571–578

Feldkotter M, Schwarzer V, Wirth R et al (2002) Quantitative analyses of SMN1 and SMN2 based on real-time lightCycler PCR: fast and highly reliable carrier testing and prediction of severity of spinal muscular atrophy. Am J Hum Genet 70:358–368

Finsterer J (2005) Mitochondrial neuropathy. Clin Neurol Neurosurg 107:181–186

Finsterer J (2011) Inherited mitochondrial neuropathies. J Neurol Sci 304:9–16

Finsterer J, Scorza FA (2017) Effects of antiepileptic drugs on mitochondrial functions, morphology, kinetics, biogenesis, and survival. Epilepsy Res 136:5–11

Finsterer J, Zarrouk-Mahjoub S (2016) Treatment of muscle weakness in neuromuscular disorders. Expert Rev Neurother 16:1383–1395

Fischbach KF, Heisenberg M (1981) Structural brain mutant of *Drosophila* melanogaster with reduced cell numbers in the medulla cortex and with normal optomotor yaw response. Proc Natl Acad Sci U S A 78:1105–1109

Fortier LA, Travis AJ (2011) Stem cells in veterinary medicine. Stem Cell Res Ther 2:9

Franco B, Bogdanik L, Bobinnec Y et al (2004) Shaggy, the homolog of glycogen synthase kinase 3, controls neuromuscular junction growth in *Drosophila*. J Neurosci 24:6573–6577

Frisbie DD, Smith RK (2010) Clinical update on the use of mesenchymal stem cells in equine orthopaedics. Equine Vet J 42:86–89

Fujita R, Tamai K, Aikawa E et al (2015) Endogenous mesenchymal stromal cells in bone marrow are required to preserve muscle function in mdx mice. Stem Cells 33:962–975

Galli R, Borello U, Gritti A et al (2000) Skeletal myogenic potential of human and mouse neural stem cells. Nat Neurosci 3:986–991

Gariepy CE (2004) Developmental disorders of the enteric nervous system: genetic and molecular bases. J Pediatr Gastroenterol Nutr 39:5–11

Garne E, Dolk H, Krageloh-Mann I et al (2008) Cerebral palsy and congenital malformations. Eur J Paediatr Neurol 12:82–88

Genova E, Pelin M, Sasaki K et al (2018) Induced pluripotent stem cells as a model for therapy personalization of pediatric patients: disease modeling and drug adverse effects prevention. Curr Med Chem 25(24):2826–2839. https://doi.org/10.2174/0929867324666170804150131

Gerlinger M, Rowan AJ, Horswell S (2012) Intratumor heterogeneity and branched evolution revealed by multiregion sequencing. N Engl J Med 366:883–892

Gershon MD, Ratcliffe EM (2004) Developmental biology of the enteric nervous system: pathogenesis of Hirschsprung's disease and other congenital dysmotilities. Semin Pediatr Surg 13:224–235

Goldman SA (2016) Stem and progenitor cell-based therapy of the central nervous system: hopes, hype, and wishful thinking. Cell Stem Cell 18:174–188

Gonçalves MA, de Vries AA, Holkers M et al (2006) Human mesenchymal stem cells ectopically expressing full-length dystrophin can complement Duchenne muscular dystrophy myotubes by cell fusion. Hum Mol Genet 15:213–221

Grill R, Murai K, Blesch A et al (1997) Cellular delivery of neurotrophin-3 promotes corticospinal axonal growth and partial functional recovery after spinal cord injury. J Neurosci 17:5560–5572

Gupta S, Kulhara P (2010) What is schizophrenia: a neurodevelopmental or neurodegenerative disorder or a combination of both? A critical analysis. Indian J Psychiatry 52:21–27

Gutierrez-Nibeyro SD (2011) Commercial cell-based therapies for musculoskeletal injuries in horses. Vet Clin North Am Equine Pract 27:363–371

Hall MN, Hall JK, Cadwallader AB et al (2017) Transplantation of skeletal muscle stem cells. Methods Mol Biol 1556:237–244

Hamid S, Ray H (2008) Role of electrical stimulation for rehabilitation and regeneration after spinal cord injury: an overview. Eur Spine J 17:1256–1269

Hardy J, Gwinn-Hardy K (1998) Genetic classification of primary neurodegenerative disease. Science 282:1075–1079

Hassani A, Horvath R, Chinnery PF (2010) Mitochondrial myopathies: developments in treatment. Curr Opin Neurol 23:459–465

Hatten ME, Kettenmann H, Ransom BR (1991) Glial cell lineage. Glia 4:124–243

Heilker R, Traub S, Reinhardt P et al (2014) iPS cell derived neuronal cells for drug discovery. Trends Pharmacol Sci 35:510–519

Hermann A, Storch A (2013) Induced neural stem cells (iNSCs) in neurodegenerative diseases. J Neural Transm 120(Suppl 1):S19–S25

Hermann A, Gastl R, Liebau S et al (2004) Efficient generation of neural stem cell-like cells from adult human bone marrow stromal cells. J Cell Sci 117:4411–4422

Hermann A, Maisel M, Wegner F et al (2006a) Multipotent neural stem cells from the adult tegmentum with dopaminergic potential develop essential properties of functional neurons. Stem Cells 24:949–964

Hermann A, Maisel M, Storch A (2006b) Epigenetic conversion of human adult bone mesodermal stromal cells into neuroectodermal cell types for replacement therapy of neurodegenerative disorders. Expert Opin Biol Ther 6:653–670

Hessinger C, Technau GM, Rogulja-Ortmann A (2017) The *Drosophila* Hox gene *Ultrabithorax* acts in both muscles and motoneurons to orchestrate formation of specific neuromuscular connections. Development 144:139–150

Hiemer SE, Varelas X (2013) Stem cell regulation by the Hippo pathway. Biochim Biophys Acta 1830:2323–2334

Hillger JM, Lieuw WL, Heitman LH et al (2017) Label-free technology and patient cells: from early drug development to precision medicine. Drug Discov Today 22:1808–1815

Hilt RJ, Chaudhari M, Bell JF et al (2014) Side effects from use of one or more psychiatric medications in a population-based sample of children and adolescents. J Child Adolesc Psychopharmacol 24:83–89

Hirsch NP (2007) Neuromuscular junction in health and disease. Br J Anaesth 99:132–138

Hodgson B, Mafi R, Mafi P et al (2018) The regulation of differentiation of mesenchymal stem-cells into skeletal muscle: a look at signalling molecules involved in myogenesis. Curr Stem Cell Res Ther 13(5):384–407. https://doi.org/10.2174/1574888X11666170907113151

Holland EC (2000) Glioblastoma multiforme: the terminator. Proc Natl Acad Sci USA 97:6242–6244

Horner PJ, Fred HG (2000) Regenerating the damaged central nervous system. Nature 407:963–970

Hou XQ, Wang L, Wang FG et al (2017) Combination of RNA interference and stem cells for treatment of central nervous system diseases. Genes (Basel) 8:135. https://doi.org/10.3390/genes8050135

Hryhorowicz S, Walczak M, Zakerska-Banaszak O et al (2018) Pharmacogenetics of cannabinoids. Eur J Drug Metab Pharmacokinet 43(1):1–12. https://doi.org/10.1007/s13318-017-0416-z

Hu BY, Weick JP, Yu J, Ma LX et al (2010) Neural differentiation of human induced pluripotent stem cells follows developmental principles but with variable potency. Proc Natl Acad Sci U S A 107:4335–4340

Huang HI, Shih R (2015) Neurotropic enterovirus infections in the central nervous system. Viruses 7:6051–6066

Ibi D, de la Feunte Revenga M, Kezunovic N et al (2017) Antipsychotic-induced Hdac2 transcription via NF-κB leads to synaptic and cognitive side effects. Nat Neurosci 20:1247–1259

Ichim TE, Solano F, Glenn E et al (2007) Stem cell therapy for autism. J Transl Med 5:30

Imitola J, Raddassi K, Park KI et al (2004) Directed migration of neural stem cells to sites of CNS injury by the stromal cell-derived factor 1alpha/CXC chemokine receptor 4 pathway. Proc Natl Acad Sci U S A 101:18117–18122

Ishiuchi T, Torres-Padilla ME (2013) Towards an understanding of the regulatory mechanisms of totipotency. Curr Opin Genet Dev 23:512–518

Islam MN, Das SR, Emin MT et al (2012) Mitochondrial transfer from bone-marrow-derived stromal cells to pulmonary alveoli protects against acute lung injury. Nat Med 18:759–765

Jackson JS, Golding JP, Chapon C et al (2010) Homing of stem cells to sites of inflammatory brain injury after intracerebral and intravenous administration: a longitudinal imaging study. Stem Cell Res Ther 1:17. https://doi.org/10.1186/scrt17

Jaeger J, Martinez-Arias A (2009) Getting the measure of positional information. PLoS Biol 7:e1000081

Jang J, Yoo JE, Lee JA et al (2012) Disease-specific induced pluripotent stem cells: a platform for human disease modeling and drug discovery. Exp Mol Med 44:202–213

Jarmalaviciute A, Tunaitis V, Pivoraite U et al (2015) Exosomes from dental pulp stem cells rescue human dopaminergic neurons from 6-hydroxy-dopamine-induced apoptosis. Cytotherapy 17:932–939

Jarvinen TA, Järvinen TL, Kaariainen M et al (2005) Muscle injuries: biology and treatment. Am J Sports Med 33:745–764

Jarvinen TA, Jarvinen TL, Kaariainen M et al (2007) Muscle injuries: optimizing recovery. Best Pract Res Clin Rheumatol 21:317–331

Jarvinen TA, Jarvinen M, Kalimo H (2014) Regeneration of injured skeletal muscle after the injury. Muscles Ligaments Tendons J 3:337–345

Jarvis S, Glinianaia SV, Torrioli MG et al (2003) Cerebral palsy and intrauterine growth in single births: European collaborative study. Lancet 362:1106–1111

Jarvis S, Glinianaia SV, Blair E (2006) Cerebral palsy and intrauterine growth. Clin Perinatal 33:285–300

Javitt DC, Spencer KM, Thaker GK et al (2008) Neurophysiological biomarkers for drug development in schizophrenia. Nature Rev Drug Discov 7:68–83

Jerath NU, Shy ME (2015) Hereditary motor and sensory neuropathies: understanding molecular pathogenesis could lead to future treatment strategies. Biochim Biophys Acta 1852:667–678

Ji JF, He BP, Dheen ST et al (2004) Interactions of chemokines and chemokine receptors mediate the migration of mesenchymal stem cells to the impaired site in the brain after hypoglossal nerve injury. Stem Cells 22:415–427

Jiang D, Gao F, Zhang Y et al (2016) Mitochondrial transfer of mesenchymal stem cells effectively protects corneal epithelial cells from mitochondrial damage. Cell Death Dis 7:e2467. https://doi.org/10.1038/cddis.2016.358

Johnson RT (1984) Slow viral infections of the nervous system. J Infect Dis Antimicrob Agents 1:85–89

Johnson BE, Mazor T, Hong C (2014) Mutational analysis reveals the origin and therapy-driven evolution of recurrent glioma. Science 343:189–193

Jones DL, Wagers AJ (2008) No place like home: anatomy and function of the stem cell niche. Nat Rev Mol Cell Biol 9:11–21

Joo WJ, Sweeney LB, Liang L et al (2013) Linking cell fate, trajectory choice, and target selection: genetic analysis of Sema-2b in olfactory axon targeting. Neuron 78:673–686

Jopling C, Boue S, Izpisua Belmonte JC (2011) Dedifferentiation, transdifferentiation and reprogramming: three routes to regeneration. Nat Rev Mol Cell Biol 12:79–89
Jubelt B (1992) Motor neuron diseases and viruses: poliovirus, retroviruses, and lymphomas. Curr Opin Neurol Neurosurg 5:655–658
Jubelt B, Cashman NR (1987) Neurological manifestations of the post-polio syndrome. Crit Rev Neurobiol 3:199–220
Jubelt B, Drucker J (1999) Poliomyelitis and the post-polio syndrome. In: Younger DS (ed) Motor disorders. Williams & Wilkins, Philadelphia, pp 381–395
Kabashi E, Brustein E, Champagne N et al (2011) Zebrafish models for the functional genomics of neurogenetic disorders. Biochim Biophys Acta 1812:335–345
Kallur T, Darsalia V, Lindvall O et al (2006) Human fetal cortical and striatal neural stem cells generate region-specific neurons in vitro and differentiate extensively to neurons after intrastriatal transplantation in neonatal rats. J Neurosci Res 84:1630–1644
Kang JF, Tang BS, Guo JF (2016a) The progress of induced pluripotent stem cells as models of Parkinson's disease. Stem Cells Int 2016:4126214. https://doi.org/10.1155/2016/4126214
Kang JM, Yeon BK, Cho SJ et al (2016b) Stem cell therapy for Alzheimer's disease: a review of recent clinical trials. J Alzheimers Dis 54:879–889
Kao HT, Ryoo K, Lin A et al (2017) Synapsins regulate brain-derived neurotrophic factor-mediated synaptic potentiation and axon elongation by acting on membrane rafts. Eur J Neurosci 45:1085–1101
Katsuda T, Tsuchiya R, Kosaka N et al (2013) Human adipose tissue-derived mesenchymal stem cells secrete functional neprilysin-bound exosomes. Sci Rep 3:1197
Keshishian H, Chiba A (1993) Neuromuscular development in *Drosophila*: insights from single neurons and single genes. Trends Neurosci 16:278–283
Kiernan MC, Vucic S, Cheah BC et al (2011) Amyotrophic lateral sclerosis. Lancet 377:942–955
Kim YJ, Lee G (2013) Candidate ALS therapeutics motor toward "in vitro clinical trials". Cell Stem Cell 12:633–634
Kim GM, Xu J, Xu J et al (2001) Tumor necrosis factor receptor deletion reduces nuclear factor-kappaB activation, cellular inhibitor of apoptosis protein 2 expression, and functional recovery after traumatic spinal cord injury. J Neurosci 21:6617–6625
Kim H, Cooke MJ, Shoichet MS (2012) Creating permissive microenvironments for stem cell transplantation into the central nervous system. Trends Biotech 30:55–63
Kley RA, Tarnopolsky MA, Vorgerd M (2008) Creatine treatment in muscle disorders: a meta-analysis of randomised controlled trials. J Neurol Neurosurg Psychiatry 79:366–367
Kmet M, Guo C, Edmondson C et al (2013) Directed differentiation of human embryonic stem cells into corticofugal neurons uncovers heterogeneous Fezf2-expressing subpopulations. PLoS One 8:e67292
Ko HC, Gelb BD (2014) Concise review: drug discovery in the age of the induced pluripotent stem cell. Stem Cells Transl Med 3:500–509
Kokontis L, Gutmann L (2000) Current treatment of neuromuscular diseases. Arch Neurol 57:939–943
Koon AC, Chan HYE (2017) *Drosophila* melanogaster as a model organism to study RNA toxicity of repeat expansion-associated neurodegenerative and neuromuscular diseases. Front Cell Neurosci 11:1–22
Koyuncu OO, Hogue IB, Enquist LW (2013) Virus infections in the nervous system. Cell Host Microbe 13:379–393
Krabbe C, Zimmer J, Meyer M (2005) Neural transdifferentiation of mesenchymal stem cells-a critical review. APMIS 113:831–844
Kuzdas-Wood D, Stefanova N, Jellinger KA et al (2014) Towards translational therapies for multiple system atrophy. Prog Neurobiol 118:19–35
Lacorazza HD, Flax JD, Snyder EY et al (1996) Expression of human beta-hexosaminidase alpha-subunit gene (the gene defect of Tay-Sachs disease) in mouse brains upon engraftment of transduced progenitor cells. Nat Med 2:424–429

Lai RC, Yeo RW, Lim SK (2015) Mesenchymal stem cell exosomes. Semin Cell Dev Biol 40:82–88
Laing NG (2012) Genetics of neuromuscular disorders. Crit Rev Clin Lab Sci 49:33–48
Lake JI, Heuckeroth RO (2013) Enteric nervous system development: migration, differentiation, and disease. Am J Physiol Gastrointest Liver Physiol 305:G1–G24
Lala-Tabbert N, Fu D, Wiper-Bergeron N (2016) Induction of CCAAT/Enhancer-Binding protein beta expression with the phosphodiesterase inhibitor Isobutylmethylxanthine improves myoblast engraftment into dystrophic muscle. Stem Cells Transl Med 5:500–510
Lamfers M, Idema S, van Milligen F et al (2009) Homing properties of adipose-derived stem cells to intracerebral glioma and the effects of adenovirus infection. Cancer Lett 274:78–87
Lane SW, Scadden DT, Gilliland DG (2009) The leukemic stem cell niche: current concepts and therapeutic opportunities. Blood 114:1150–1157
Lane SW, Williams DA, Watt FM (2014) Modulating the stem cell niche for tissue regeneration. Nat Biotechnol 32:795–803
Lange W (1975) Cell number and cell density in the cerebellar cortex of man and other mammals. Cell Tiss Res 157:115–124
Lara-Velazquez M, Akinduroa OO, Reimera R et al (2017) Stem cell therapy and its potential role in pituitary disorders. Neuroendocrinology 24:1–9
Lax NZ, Hepplewhite PD, Reeve AK et al (2012) Cerebellar ataxia in patients with mitochondrial DNA disease: a molecular clinicopathological study. J Neuropathol Exp Neurol 71:148–161
Lax NZ, Gorman GS, Turnbull DM (2017) Central nervous system involvement in mitochondrial disease. Neuropathol Appl Neurobiol 43:102–118
Lee SK, Pfaff SL (2001) Transcriptional networks regulating neuronal identity in the developing spinal cord. Nat Neurosci Suppl 4:1183–1191
Lee J, Kuroda S, Shichinohe H et al (2003) Migration and differentiation of nuclear fluorescence-labeled bone marrow stromal cells after transplantation into cerebral infarct and spinal cord injury in mice. Neuropathology 23:169–180
Lee M, Liu T, Im W et al (2016) Exosomes from adipose-derived stem cells ameliorate phenotype of Huntington's disease in vitro model. Eur J Neurosci 44:2114–2119
Lefebvre S, Bürglen L, Reboullet S et al (1995) Identification and characterization of a spinal muscular atrophy-determining gene. Cell 80:155–165
Lefebvre S, Burlet P, Liu Q et al (1997) Correlation between severity and SMN protein level in spinal muscular atrophy. Nat Genet 16:265–269
Leibacher J, Henschler R (2016) Biodistribution, migration and homing of systemically applied mesenchymal stem/stromal cells. Stem Cell Res Ther 7:7
Li XJ, Du ZW, Zarnowska ED et al (2005) Specification of motoneurons from human embryonic stem cells. Nat Biotechnol 23:215–221
Li Y, Wu WH, Hsu CW et al (2014) Gene therapy in patient-specific stem cell lines and a preclinical model of retinitis pigmentosa with membrane frizzled-related protein defects. Mol Ther 22:1688–1697
Li H, Saucedo-Cuevas L, Regla-Nava JA et al (2016a) Zika virus infects neural progenitors in the adult mouse brain and alters proliferation cell. Stem Cell 19:593–598
Li Y, Chan L, Nguyen HV, Tsang SH (2016b) Personalized medicine: cell and gene therapy based on patient-specific iPSC-derived retinal pigment epithelium cells. Adv Exp Med Biol 854:549–555
Lin H, Li Q, Lei Y (2017) An integrated miniature bioprocessing for personalized human induced pluripotent stem cell expansion and differentiation into neural stem cells. Sci Rep 7:40191. https://doi.org/10.1038/srep40191
Liu Y, Deng W (2016) Reverse engineering human neurodegenerative disease using pluripotent stem cell technology. Brain Res 1638:30–41
Liu L, Rando TA (2011) Manifestations and mechanisms of stem cell aging. J Cell Biol 193:257–266
Liu Y, Zhang SC (2010) Human stem cells as a model of motoneuron development and diseases. Ann N Y Acad Sci 1198:192–200

Liu H, Jiang D, Chi F et al (2012) The Hippo pathway regulates stem cell proliferation, self-renewal, and differentiation. Protein Cell 3:291–304
Liu Y, Weick JP, Liu H et al (2013) Medial ganglionic eminence-like cells derived from human embryonic stem cells correct learning and memory deficits. Nat Biotechnol 31:440–447
Lloyd TE, Taylor JP (2010) Flightless Flies: *Drosophila* models of neuromuscular disease. Ann N Y Acad Sci 1184:e1–e20
Lloyd-Lewis B, Harris OB, Watson CJ et al (2017) Mammary stem cells: premise, properties, and perspectives. Trends Cell Biol 27:556–567
Lnenicka GA, Keshishian H (2000) Identified motor terminals in *Drosophila* larvae show distinct differences in morphology and physiology. J Neurobiol 43:186–197
Lorant J, Saury C, Schleder C et al (2018) Skeletal muscle regenerative potential of human MuStem cells following transplantation into injured mice muscle. Mol Ther 26(2):618–633. https://doi.org/10.1016/j.ymthe.2017.10.013
Lorenzoni PJ, Scola RH, Kay CSK et al (2012) Congenital myasthenic syndrome: a brief review. Pediatric Neurol 46:141–148
Lossing A, Moore M, Zuhl M (2017) Dance as a treatment for neurological disorders. Body Mov Dance Psychother 12:170–184
Loy JH, Merry SN, Hetrick SE et al (2017) Atypical antipsychotics for disruptive behaviour disorders in children and youths. Cochrane Database of Syst Rev 9:CD008559. https://doi.org/10.1002/14651858.CD008559.pub2
Lu F, Zhang Y (2015) Cell totipotency: molecular features, induction, and maintenance. Natl Sci Rev 2:217–225
Lu A, Poddar M, Tang Y et al (2014) Rapid depletion of muscle progenitor cells in dystrophic mdx/utrophin2/2 mice. Hum Mol Genet 3:4786–4800
Lyons MK (2011) Deep brain stimulation: current and future clinical applications. Mayo Clin Proc 86:662–672
Ma H, Folmes CD, Wu J et al (2015) Metabolic rescue in pluripotent cells from patients with mtDNA disease. Nature 524:234–238
Mackay-Sim A (2013) Patient-derived stem cells: pathways to drug discovery for brain diseases. Front Cell Neurosci 7:29. https://doi.org/10.3389/fncel.2013.00029
Maclean S, Khan WS, Malik AA et al (2012) The potential of stem cells in the treatment of skeletal muscle injury and disease. Stem Cells Int 2012. https://doi.org/10.1155/2012/282348
Mafi R, Hindocha S, Mafi P et al (2011) Sources of adult mesenchymal stem cells applicable for musculoskeletal applications—a systematic review of the literature. Open Orthop J 5(Suppl 2):242–248
Maherali N, Hochedlinger K (2008) Induced pluripotency of mouse and human somatic cells. Cold Spring Harb Symp Quant Biol 73:157–162
Maherali N, Sridharan R, Xie W et al (2007) Directly reprogrammed fibroblasts show global epigenetic remodeling and widespread tissue contribution. Cell Stem Cell 1:55–70
Mali P, Esvelt KM, Church GM (2013) Cas9 as a versatile tool for engineering biology. Nat Methods 10:957–963
Malone DA, Pandya MM (2006) Behavioral neurosurgery. Adv Neurol 99:241–247
Manzanares M, Rodriguez TA (2013) Development: Hippo signalling turns the embryo inside out. Curr Biol 23:R559–R561
Marano G, Traversi G, Romagnoli E (2011) Cardiologic side effects of psychotropic drugs. Am J Geriatr Cardiol 8:243–253
Marder E (2002) Non-mammalian models for studying neural development and function. Nature 417:318–321
Maumus M, Jorgensen C, Noel D (2013) Mesenchymal stem cells in regenerative medicine applied to rheumatic disease: role of secretome and exosomes. Biochimie 95:2229–2234
Mauro A (1961) Satellite cell of skeletal muscle fibers. J Biophys Biochem Cytol 9:493–495
Mayberg HS, Lozano AM, Voon V (2005) Deep brain stimulation for treatment-resistant depression. Neuron 45:651–660

McConnell SK (1995) The generation of neuronal diversity in the developing central nervous system. J Neurosci 15:6987–6998
McTigue DM, Horner PJ, Stokes BT, Gage FH (1998) Neurotrophin-3 and brain-derived neurotrophic factor induce oligodendrocyte proliferation and myelination of regenerating axons in the contused adult rat spinal cord. J Neurosci 18:5354–5365
McVey KA, Mink JA, Snapp IB et al (2012) *Caenorhabditis elegans*: an emerging model system for pesticide neurotoxicity. J Environ Anal Toxicol S4:003. https://doi.org/10.4172/2161-0525.S4-003
Meadow LA, Gell D, Broadie K et al (1994) The cell adhesion molecule, connectin, and the development of the *Drosophila* neuromuscular system. J Cell Sci 107:321–328
Mendelsohn AR, Larrick JW (2015) Stem cell depletion by global disorganization of the H3K9me3 epigenetic marker in aging. Rejuvenation Res 18:371–375
Menon KP, Carrillo RA, Zinn K (2013) Development and plasticity of the *Drosophila* larval neuromuscular junction. Wiley Interdiscip Rev Dev Biol 2:647–670
Metzger M, Caldwell C, Barlow AJ et al (2009a) Enteric nervous system stem cells derived from human gut mucosa for the treatment of aganglionic gut disorders. Gastroenterology 136:2214–2225
Metzger M, Bareiss PM, Danker T et al (2009b) Expansion and differentiation of neural progenitors derived from the human adult enteric nervous system. Gastroenterology 137:2063–2073
Miale IL, Sidman RL (1961) An autoradiographic analysis of histogenesis in the mouse cerebellum. Exp Neurol 4:277–296
Mian MK, Campos M, Sheth S et al (2010) Deep brain stimulation for obsessive-compulsive disorder: past, present, and future. J Neurosurg 29:E10
Mizuno H, Zuk PA, Zhu M et al (2002) Myogenic differentiation by human processed lipoaspirate cells. Plast Reconstr Surg 109:199–209
Moavero R, Santarone ME, Galasso C et al (2017) Cognitive and behavioral effects of new antiepileptic drugs in pediatric epilepsy. Brain Dev 39:464–469
Moller AR (2001) Symptoms and signs caused by neural plasticity. Neurol Res 23:565–572
Montarras D, Morgan J, Collins C et al (2005) Direct isolation of satellite cells for skeletal muscle regeneration. Science 309:2064–2067
Moore XL, Lu J, Sun L et al (2004) Endothelial progenitor cells' "homing" specificity to brain tumors. Gene Ther 1:811–818
Morgani SM, Brickman JM (2014) The molecular underpinnings of totipotency. Philos Trans R Soc Lond B Biol Sci 369(1657):pii: 20130549. https://doi.org/10.1098/rstb.2013.0549
Morrison SJ, Kimble J (2006) Asymmetric and symmetric stem-cell divisions in development and cancer. Nature 441:1068–1074
Morrison JI, Loof S, He P et al (2006) Salamander limb regeneration involves the activation of a multipotent skeletal muscle satellite cell population. J Cell Biol 172:433–440
Morshead CM, Reynolds BA, Craig CG et al (1994) Neural stem cells in the adult mammalian forebrain: a relatively quiescent subpopulation of subependymal cells. Neuron 13:1071–1082
Muench J, Hamer AM (2010) Adverse effects of antipsychotic medications. Am Fam Physician 81:617–622
Muntoni F, Brown S, Sewry C et al (2002) Muscle development genes: their relevance in neuromuscular disorders. Neuromuscul Disord 12:438–446
Namiki J, Kojima A, Tator CH (2000) Effect of brain-derived neurotrophic factor, nerve growth factor, and neurotrophin-3 on functional recovery and regeneration after spinal cord injury in adult rats. J Neurotrauma 17:1219–1231
Narayanan K, Mishra S, Singh S et al (2017) Engineering concepts in stem cell research. Biotechnol J 12. https://doi.org/10.1002/biot.201700066
Narita Y, Rijli FM (2009) *Hox* genes in neural patterning and circuit formation in the mouse hindbrain. Curr Top Dev Biol 88:139–167
Nathanson N (2008) The pathogenesis of poliomyelitis: what we don't know. Adv Virus Res 71:1–50

Naya FJ, Olson E (1999) MEF2: a transcriptional target for signalling pathways controlling skeletal muscle growth and differentiation. Curr Opin Cell Biol 11:683–688
Nelson CM (2009) Geometric control of tissue morphogenesis. Biochim Biophys Acta 1793:903–910
Neupane M, Chang CC, Kiupel M et al (2008) Isolation and characterization of canine adipose-derived mesenchymal stem cells. Tissue Eng Part A 14:1007–1015
Newgreen D, Young HM (2002a) Enteric nervous system: development and developmental disturbances—part 1. Pediatr Dev Pathol 5:224–247
Newgreen D, Young HM (2002b) Enteric nervous system: development and developmental disturbances—part 2. Pediatr Dev Pathol 5:329–349
Nicolson GL (2008) Chronic bacterial and viral infections in neurodegenerative and neurobehavioral diseases. Labmedicine 39:291–299
Nieto MA (1996) Molecular biology of axon guidance. Neuron 17:1039–1048
Nishio Y, Koda M, Hashimoto M et al (2009) Deletion of macrophage migration inhibitory factor attenuates neuronal death and promotes functional recovery after compression-induced spinal cord injury in mice. Acta Neuropathol 117:321–328
Noyola DE, Jimenez-Capdeville ME, Demmler-Harrison GJ (2010) Central nervous system disorders in infants with congenital cytomegalovirus infection. Neurol Res 32:278–284
O'Kane CJ (2011) *Drosophila* as a model organism for the study of neuropsychiatric disorders. In: Hagan JJ (ed) Molecular and functional models in neuropsychiatry. Current topics in behavioral neurosciences, vol 7. Springer, Berlin, pp 37–60
Obermair FJ, Schroter A, Thallmair M (2008) Endogenous neural progenitor cells as therapeutic target after spinal cord injury. Physiology 23:296–304
Okita K, Ichisaka T, Yamanaka S (2007) Generation of germline-competent induced pluripotent stem cells. Nature 448:313–317
Orbay H, Mizuno H (2013) Translational research in stem cell treatment of neuromuscular diseases. ISRN Stem Cells 2013:947329. https://doi.org/10.1155/2013/947329
Ottoboni L, Merlini A, Martino G (2017) Neural stem cell plasticity: advantages in therapy for the injured central nervous system. Front Cell Dev Biol 5:52
Page RL, Malcuit C, Vilner L et al (2011) Restoration of skeletal muscle defects with adult human cells delivered on fibrin microthreads. Tissue Eng Part A 17:2629–2640
Palmer TD, Takahashi J, Gage FH (1997) The adult rat hippocampus contains primordial neural stem cells. Mol Cell Neurosci 8:389–404
Pandya H, Shen MJ, Ichikawa DM et al (2017) Differentiation of human and murine induced pluripotent stem cells to microglia-like cells. Nat Neurosci 20:753–759
Pannerec A, Marazzi G, Sassoon D (2012) Stem cells in the hood: the skeletal muscle niche. Trends Mol Med 18:599–606
Pareyson D, Piscosquito G, Moroni I et al (2013) Peripheral neuropathy in mitochondrial disorders. Lancet Neurol 12:1011–1024
Parisi MA, Kapur RP (2000) Genetics of Hirschsprung disease. Curr Opin Pediatr 12:610–617
Park IH, Arora N, Huo H et al (2008) Disease-specific induced pluripotent stem cells. Cell 134:877–886
Parr AM, Kulbatski I, Tator CH (2007) Transplantation of adult rat spinal cord stem/progenitor cells for spinal cord injury. J Neurotrauma 24:835–845
Patten SA, Armstrong GA, Lissouba A et al (2014) Fishing for causes and cures of motor neuron disorders. Dis Model Mech 7:799–809
Peerani R, Zandstra PW (2010) Enabling stem cell therapies through synthetic stem cell-niche engineering. J Clin Invest 120:60–70
Pendleton C, Li Q, Chesler DA et al (2013) Mesenchymal stem cells derived from adipose tissue vs bone marrow: in vitro comparison of their tropism towards gliomas. PLoS One 8:e58198
Penzo-Mendez AI, Stanger BZ (2015) Organ-size regulation in mammals. Cold Spring Harb Perspect Biol 7:a019240
Pera MF, Reubinoff B, Trounson A (2000) Human embryonic stem cells. J Cell Sci 113:5–10

Perry RL, Rudnick MA (2000) Molecular mechanisms regulating myogenic determination and differentiation. Front Biosci 5:750–767
Pfeffer G, Chinnery PF (2013) Diagnosis and treatment of mitochondrial myopathies. Ann Med 45:4–16
Philippidou P, Dasen JS (2013) Hox genes: choreographers in neural development, architects of circuit organization. Neuron 80:12–34
Piccin D, Morshead CM (2010) Potential and pitfalls of stem cell therapy in old age. Dis Model Mech 3:421–425
Pierret C, Morrison JA, Kirk MD (2008) Treatment of lysosomal storage disorders: Focus on the neuronal ceroid-lipofuscinoses. Acta Neurobiol Exp 68:429–442
Pini V, Morgan JE, Muntoni F et al (2017) Genome editing and muscle stem cells as a therapeutic tool for muscular dystrophies. Curr Stem Cell Rep 3:137–148
Pisciotta A, Riccio M, Carnevale G et al (2015) Stem cells isolated from human dental pulp and amniotic fluid improve skeletal muscle histopathology in *mdx*/SCID mice. Stem Cell Res Ther 6:156
Plantie E, Migocka-Patrzałek M, Daczewska M et al (2015) Model organisms in the fight against muscular dystrophy: lessons from *Drosophila* and zebrafish. Molecules 20:6237–6253
Platt FM, Lachmann RH (2009) Treating lysosomal storage disorders: current practice and future prospects. Biochim Biophys Acta 1793:737–745
Pluchino S, Quattrini A, Brambilla E et al (2003) Injection of adult neurospheres induces recovery in a chronic model of multiple sclerosis. Nature 422:688–694
Pluchino S, Gritti A, Blezer E et al (2009) Human neural stem cells ameliorate autoimmune encephalomyelitis in non-human primates. Ann Neurol 66:343–354
Poe LB, Coleman LL, Mahmud F (1989) Congenital central nervous system anomalies. RadioGraphics 9:801–826
Price DL, Sisodia SS, Borchelt DR (1998) Genetic neurodegenerative diseases: the human illness and transgenic models. Science 282:1079–1083
Prior TW (2010) Spinal muscular atrophy: a time for screening. Curr Opin Pediatr 22(6):696–702
Quigley HA, Addicks EM, Green WG (1982) Optic nerve damage in human glaucoma. III. Quantitative correlation of nerve fiber loss and visual field defect in glaucoma, ischemic neuropathy, papilledema, and toxic neuropathy. Arch Ophthalmol 100:135–146
Rajput BS, Chakrabarti SK, Dongare VS et al (2015) Human umbilical cord mesenchymal stem cells in the treatment of Duchenne Muscular Dystrophy: safety and feasibility study in India. J Stem Cells 10:141–156
Rakic P (1985) Limits of neurogenesis in primates. Science 227:1054–1056
Ramer MS, John VP, Stephen BM (2000) Functional regeneration of sensory axons into the adult spinal cord. Nature 403:312–316
Ramos A, Camargo FD (2012) The Hippo signaling pathway and stem cell biology. Trends Cell Biol 22:339–346
Redondo PA, Pavlou M, Loizidou M et al (2017) Elements of the niche for adult stem cell expansion. J Tissue Eng 8. https://doi.org/10.1177/2041731417725464
Regal M, Paramo C, Sierra SM et al (2001) Prevalence and incidence of hypopituitarism in an adult Caucasian population in northwestern Spain. Clin Endocrinol 55:73
Reif A, Fritzen S, Finger M et al (2006) Neural stem cell proliferation is decreased in schizophrenia, but not in depression. Mol Psychiatry 11:514–522
Relaix F, Rocancourt D, Mansouri A et al (2005) A Pax3/Pax7-dependent population of skeletal muscle progenitor cells. Nature 435:948–953
Renton AE, Chio A, Traynor BJ (2014) State of play in amyotrophic lateral sclerosis genetics. Nat Neurosci 17:17–23
Robinton DA, Daley GQ (2012) The promise of induced pluripotent stem cells in research and therapy. Nature 481:295–305
Rohwedel J, Maltsev V, Bober E et al (1994) Muscle cell differentiation of embryonic stem cells reflects myogenesis in vivo: developmentally regulated expression of myogenic determination genes and functional expression of ionic currents. Dev Biol 164:87–101

Rosengardten Y, McKenna T, Grochova D et al (2011) Stem cell depletion in Hutchinson–Gilford progeria syndrome. Aging Cell 10:1011–1020
Ross CA, Tabrizi SJ (2010) Huntington's disease: from molecular pathogenesis to clinical treatment. Lancet Neurol 10:83–98
Rossi CA, Flaibani M, Blaauw B et al (2011) In vivo tissue engineering of functional skeletal muscle by freshly isolated satellite cells embedded in a photopolymerizable hydrogel. FASEB J 25:2296–2304
Roux M, Zaffran S (2016) Hox genes in cardiovascular development and diseases. J Dev Biol 4:14
Ruff CA, Fehlings MG (2010) Neural stem cells in regenerative medicine: bridging the gap. Panminerva Med 52:125–147
Ruller CM, Tabor-Godwin JM, Van Deren Jr DA et al (2012) Neural stem cell depletion and CNS developmental defects after enteroviral infection. Am J Pathol 180:1107–1120
Ruven C, Li W, Li H et al (2017) Transplantation of embryonic spinal cord derived cells helps to prevent muscle atrophy after peripheral nerve injury. Int J Mol Sci 18:511. https://doi.org/10.3390/ijms18030511
Sabourin LA, Rudnicki MA (2000) The molecular regulation of myogenesis. Clin Genet 57:16–25
Salah-Mohellibi N, Millet G, André-Schmutz I et al (2006) Bone marrow transplantation attenuates the myopathic phenotype of a muscular mouse model of spinal muscular atrophy. Stem Cells 24:2723–2732
Salehi H, Amirpour N, Niapour A et al (2016) An overview of neural differentiation potential of human adipose derived stem cells. Stem Cell Rev 12:26–41
Scaglia F, Wong LJ, Vladutiu GD et al (2015) Predominant cerebellar volume loss as a neuroradiologic feature of pediatric respiratory chain defects. Am J Neuroradiol 26:1675–1680
Schappi MG, Staiano A, Milla PJ et al (2013) A practical guide for the diagnosis of primary enteric nervous system disorders. J Pediatr Gastroenterol Nutr 57:677–686
Schwalb JM, Hamani C (2008) The history and future of deep brain stimulation. Neurotherapeutics 5:3–13
Schwank G, Basler K (2010) Regulation of organ growth by morphogen gradients. Cold Spring Harb Perspect Biol 2:a001669
Schwartz MP, Hou Z, Propson NE et al (2015) Human pluripotent stem cell-derived neural constructs for predicting neural toxicity. Proc Natl Acad Sci U S A 112:12516–12521
Seale P, Sabourin LA, Girgis-Gabardo A et al (2000) Pax7 is required for the specification of myogenic satellite cells. Cell 102:777–786
Segraves RT (1988) Sexual side-effects of psychiatric drugs. Intl J Psychiatry Med 18:243–252
Sharifi MS (2013) Treatment of neurological and psychiatric disorders with deep brain stimulation; raising hopes and future challenges. Basic Clin Neurosci 4:266–270
Sharpless NE, DePinho RA (2007) How stem cells age and why this makes us grow old. Nat Rev Mol Biol 8:703–713
She S, Wei Q, Kang B et al (2017) Cell cycle and pluripotency: convergence on octamer-binding transcription factor 4 (review). Mol Med Rep 16:6459–6466
Shetty AK, Bates A (2016) Potential of GABA-ergic cell therapy for schizophrenia, neuropathic pain, and Alzheimer's and Parkinson's diseases. Brain Res 1638:74–87
Shi X, Garry DJ (2006) Muscle stem cells in development, regeneration, and disease. Genes & Dev 20:1692–1708
Shi Y, Kirwan P, Livesey FJ (2012a) Directed differentiation of human pluripotent stem cells to cerebral cortex neurons and neural networks. Nat Protoc 7:1836–1846
Shi Y, Kirwan P, Smith J et al (2012b) Human cerebral cortex development from pluripotent stem cells to functional excitatory synapses. Nat Neurosci 15:477–486
Short CA, Suarez-Zayas EA, Gomez TM (2016) Cell adhesion and invasion mechanisms that guide developing axons. Curr Opin Neurobiol 39:77–85
Sieburth D, Ch'ng Q, Dybbs M et al (2005) Systematic analysis of genes required for synapse structure and function. Nature 436:510–517
Sierra A, Martin-Suáreza S, Valcarcel-Martin R et al (2015) Neuronal hyperactivity accelerates depletion of neural stem cells and impairs hippocampal heurogenesis. Cell Stem Cell 16:488–503

Signer RA, Morrison SJ (2013) Mechanisms that regulate stem cell aging and life span. Cell Stem Cell 12:152–165
Simonen M, Pedersen V, Weinmann O et al (2003) Systemic deletion of the myelin-associated outgrowth inhibitor Nogo-A improves regenerative and plastic responses after spinal cord injury. Neuron 38:201–211
Slack JM (2008) Origin of stem cells in organogenesis. Science 322:1498–1501
Smith RR, Shum-Siu A, Baltzley R et al (2006) Effects of swimming on functional recovery after incomplete spinal cord injury in rats. J Neurotrauma 23:908–919
Sohn J, Lu A, Tang Y et al (2015) Activation of non-myogenic mesenchymal stem cells during the disease progression in dystrophic dystrophin/utrophin knockout mice. Hum Mol Genet 24:3814–3829
Solmi M, Murru A, Pacchiarotti I et al (2017) Safety, tolerability, and risks associated with first- and second-generation antipsychotics: a state-of-the-art clinical review. Ther Clin Risk Manag 13:757–777
Sousa BR, Parreira RC, Fonseca EA et al (2014) Human adult stem cells from diverse origins: an overview from multiparametric immunophenotyping to clinical applications. Cytometry 85:43–77
Spradling A, Drummond-Barbosa D, Kai T (2001) Stem cells find their niche. Nature 414:98–104
Srivastava D, Olson EN (2000) A genetic blueprint for cardiac development. Nature 407:221–226
Stanger BZ (2008) Organ size determination and the limits of regulation. Cell Cycle 7:318–324
Stewart AM, Braubach O, Spitsbergen J et al (2014) Zebrafish models for translational neuroscience research: from tank to bedside. Trends Neurosci 37:264–278
Sullivan PF, Kendler KS, Neale MC (2003) Schizophrenia as a complex trait: evidence from a meta-analysis of twin studies. Arch Gen Psychiatry 60:1187–1192
Swanson PA, McGavern DB (2015) Viral diseases of the central nervous system. Curr Opin Virol 11:44–54
Tajbakhsh S (2005) Skeletal muscle stem and progenitor cells: reconciling genetics and lineage. Exp Cell Res 306:364–372
Takahashi K, Yamanaka S (2006) Induction of pluripotent stem cells from mouse embryonic and adult fibroblast cultures by defined factors. Cell 126:663–676
Tamaki S, Eckert K, He D et al (2002) Engraftment of sorted/expanded human central nervous system stem cells from fetal brain. J Neurosci Res 69:976–986
Taylor MV (2002) Muscle differentiation: how two cells become one. Curr Biol 12:R224–R228
Tedesco FS, Dellavalle A, Diaz-Manera J et al (2010) Repairing skeletal muscle: regenerative potential of skeletal muscle stem cells. J Clin Invest 120:11–19
Tedesco FS, Hoshiya H, D'Antona G et al (2011) Stem cell-mediated transfer of a human artificial chromosome ameliorates muscular dystrophy. Sci Transl Med 3:96ra78. https://doi.org/10.1126/scitranslmed.3002342
Terns RM, Terns MP (2014) CRISPR-based technologies: prokaryotic defense weapons repurposed. Trends Genet 30:111–118
Tessier-Lavigne M, Goodman CS (1996) The molecular biology axon guidance. Science 274:1123–1133
Tetzlaff W, Okon EB, Karimi-Abdolrezaee S et al (2011) A systematic review of cellular transplantation therapies for spinal cord injury. J Neurotrauma 28:1611–1682
Thier M, Worsdorfer P, Lakes YB et al (2012) Direct conversion of fibroblasts into stably expandable neural stem cells. Cell Stem Cell 10:473–479
Thorley M, Malatras A, Duddy W et al (2015) Changes in communication between muscle stem cells and their environment with aging. J Neuromuscul Dis 2:205–217
Tokuzawa Y, Kaiho E, Maruyama M et al (2003) Fbx15 is a novel target of Oct3/4 but is dispensable for embryonic stem cell self-renewal and mouse development. Mol Cell Biol 23:2699–2708
Torrente Y, Camirand G, Pisati F et al (2003) Identification of a putative pathway for the muscle homing of stem cells in a muscular dystrophy model. J Cell Biol 162:511–520

Torrente Y, Gavina M, Belicchi M et al (2006) High-resolution X-ray microtomography for three-dimensional visualization of human stem cell muscle homing. FEBS Lett 580:5759–5764
Trent C, Tsung N, Horvitz HR (1983) Egg–laying defective mutants of the nematode *Caenorhabditis elegans*. Genetics 104:619–647
Tropepe V, Hitoshi S, Sirard C et al (2001) Direct neural fate specification from embryonic stem cells: a primitive mammalian neural stem cell stage acquired through a default mechanism. Neuron 30:65–78
Tsuchida T, Ensini M, Morton SB et al (1994) Topographic organisation of embryonic motor neurons defined by expression of LIM Homeobox genes. Cell 79:957–970
Tuan RS, Boland G, Tuli R (2003) Adult mesenchymal stem cells and cell-based tissue engineering. Arthritis Res Ther 5:32–45
Tueth MJ (1994) Emergencies caused by side effects of psychiatric medications. Am J Emerg Med 12:212–216
Turinetto V, Orlando L, Giachino C (2017) Induced pluripotent stem cells: advances in the quest for genetic stability during reprogramming process. Int J Mol Sci 18:1952. https://doi.org/10.3390/ijms18091952
Tweedell KS (2017) The adaptability of somatic stem cells: a review. J Stem Cells Regen Med 13:3–13
van der Knaap MS, Valk J (1988) Classification of congenital abnormalities of the CNS. Am J Neuroradiol 9:315–326
van Velthoven CT, Sheldon RA, Kavelaars A et al (2013) Mesenchymal stem cell transplantation attenuates brain injury after neonatal stroke. Stroke 44:1426–1432
Verity C, Firth H, ffrench-Constant C (2003) Congenital abnormalities of the central nervous system. J Neurol Neurosurg Psychiatry 74:i3–i8
Villanova M, Bach JR (2015) Allogeneic mesenchymal stem cell therapy outcomes for three patients with spinal muscular atrophy type 1. Am J Phys Med Rehabil 94:410–415
Vincent A (2008) Autoimmune disorders of the neuromuscular junction. Neurol India 56:305–313
Vitale I, Manic G, De Maria R et al (2017) DNA damage in stem cells. Mol Cell 66:306–319
Wagers AJ (2012) The stem cell niche in regenerative medicine. Cell Stem Cell 10:362–369
Wang YX, Dumont NA, Rudnicki MA (2014) Muscle stem cells at a glance. J Cell Sci 127:4543–4548
Wang L, Yi F, Fu L et al (2017a) CRISPR/Cas9-mediated targeted gene correction in amyotrophic lateral sclerosis patient iPSCs. Protein Cell 8:365–378
Wang Y, Yu A, Yu FX (2017b) The Hippo pathway in tissue homeostasis and regeneration. Protein Cell 8:349–359
Weisfeldt ML, Zieman SJ (2007) Advances in the prevention and treatment of cardiovascular disease. Health Aff 26:25–37
Weiss S, Dunne C, Hewson J et al (1996) Multipotent CNS stem cells are present in the adult mammalian spinal cord and ventricular neuroaxis. J Neurosci 16:7599–7609
Weissman IL (2004) Stem cells: units of development, units of regeneration, and units in evolution. Cell 100:157–168
Wells JM, Melton DA (1999) Vertebrate endoderm development. Annu Rev Cell Dev Biol 15:393–410
Wernig M, Meissner A, Foreman R et al (2007) In vitro reprogramming of fibroblasts into a pluripotent ES-cell-like state. Nature 448:318–324
Wetts R, Herrup K (1982) Interaction of granule, Purkinje and inferior olivary neurons in lurcher chimeric mice. II. Granule cell death. Brain Res 250:358–362
Wetts R, Herrup K (1983) Direct correlation between Purkinje and granule cell number in the cerebella of lurcher chimeras and wild–type mice. Dev Brain Res 10:41–44
Wise CJ, Watt DJ, Jones GE (1996) Conversion of dermal fibroblasts to a myogenic lineage is induced by a soluble factor derived from myoblasts. J Cell Biochem 61:363–374
Wolpert L (1968) The French Flag problem: a contribution to the discussion on pattern development and regulation. In: Waddington CH (ed) Towards a theoretical biology. Edinburgh University Press, Edinburgh, pp 125–133

Wong AH, Van Tol HH (2003) Schizophrenia: from phenomenology to neurobiology. Neurosci Biobehav Rev 27:269–306
Woolf CJ, Salte M (2000) Neuronal plasticity: increasing the gain of pain. Science 288:1765–1768
Wu S, Li K, Yan Y et al (2013) Intranasal delivery of neural stem cells: a CNS-specific, non-invasive cell-based therapy for experimental autoimmune encephalomyelitis. J Clin Cell Immunol 4. https://doi.org/10.4172/2155-9899.1000142
Yaffe D (1968) Retention of differentiation potentialities during prolonged cultivation of myogenic cells. Proc Natl Acad Sci U S A 61:477–483
Yao H, Gao M, Ma J et al (2015) Transdifferentiation-induced neural stem cells promote recovery of middle cerebral artery stroke rats. PLoS One 10:e0137211. https://doi.org/10.1371/journal.pone.0137211
Ye JH, Houle JD (1997) Treatment of the chronically injured spinal cord with neurotrophic factors can promote axonal regeneration from supraspinal neurons. Exp Neurol 143:70–81
Younger DS (2004) Vasculitis of the nervous system. Curr Opin Neurol 17:317–336
Yu FX, Zhao B, Guan KL (2015) Hippo pathway in organ size control, tissue homeostasis, and cancer. Cell 163:811–828
Yun K, Wold B (1996) Skeletal muscle determination and differentiation: story of a core regulatory network and its context. Curr Opin Cell Biol 8:877–889
Yuyama K, Sun H, Usuki S et al (2015) A potential function for neuronal exosomes: sequestering intracerebral amyloid-β peptide. FEBS Lett 589:84–88
Zelentsova K, Talmi Z, Abboud-Jarrous G et al (2017) Protein S regulates neural stem cell quiescence and neurogenesis. Stem Cells 35:679–693
Zett UK, Stuve O, Patejdl R (2012) Immune-mediated CNS diseases: a review on nosological classification and clinical features. Autoimmun Rev 11:167–173
Zhang Y, Zhu Y, Li Y et al (2015) Long-term engraftment of myogenic progenitors from adipose-derived stem cells and muscle regeneration in dystrophic mice. Hum Mol Genet 24:6029–6040
Zhou JY, Zhang Z, Qian GS (2016) Mesenchymal stem cells to treat diabetic neuropathy: a long and strenuous way from bench to the clinic. Cell Death Discov 2:16055. https://doi.org/10.1038/cddiscovery.2016.55
Zueter AM, Zaiter A (2015) Infectious meningitis. Clin Microbiol Newsl 37:43–51

Correction to: Stem Cells for Nerve and Muscle Repair: Harnessing Developmental Dynamics in Therapeutics

Satish Sasikumar, Ashima Bhan, and T. K. Rajendra

Correction to:
Chapter 10 in: P. V. Pham, A. El-Hashash (eds.), *Stem Cells for Cancer and Genetic Disease Treatment*, Stem Cells in Clinical Applications, https://doi.org/10.1007/978-3-319-98065-2_10

An error in the production process unfortunately led to publication of this chapter prematurely, before incorporation of the final corrections. The version supplied here has been corrected and approved by the authors.

The following additional abbreviations were added to the chapter:

Myf5	Myogenic factor 5
MyoD	Myoblast determination protein
RNA	Ribonucleic acid
PICs	$PW1^+$ interstitial cells
Sox	Sry-related High Mobility Group (HMG) box
SP	Side population
Sry	Sex determining region Y (present on the Y chromosome)
3D	Three-dimensional

The reference "Cruza PMR, Palacea J, Beeson D (2014) Congenital myasthenic syndromes and the neuromuscular junction. Curr Opin Neurol 27:566–575" was included in the chapter.

The updated online version of this chapter can be found at
https://doi.org/10.1007/978-3-319-98065-2_10

P. V. Pham, A. El-Hashash (eds.), *Stem Cells for Cancer and Genetic Disease Treatment*, Stem Cells in Clinical Applications, https://doi.org/10.1007/978-3-319-98065-2_11

The citation "Namiki et al. 2009" was corrected to read as "Namiki et al. 2000". The corresponding reference now reads as: "Namiki J, Kojima A, Tator CH (2000) Effect of brain-derived neurotrophic factor, nerve growth factor, and neurotrophin-3 on functional recovery and regeneration after spinal cord injury in adult rats. J Neurotrauma 17:1219–1231".

Index

P. V. Pham, A. El-Hashash (eds.), *Stem Cells for Cancer and Genetic Disease Treatment*, Stem Cells in Clinical Applications, https://doi.org/10.1007/978-3-319-98065-2

MIX
Papier aus verantwortungsvollen Quellen
Paper from responsible sources
FSC® C105338

If you have any concerns about our products,
you can contact us on
ProductSafety@springernature.com

In case Publisher is established outside the EU,
the EU authorized representative is:
Springer Nature Customer Service Center GmbH
Europaplatz 3, 69115 Heidelberg, Germany

Printed by Libri Plureos GmbH
in Hamburg, Germany